BUILDING
SHEDS

BUILDING SHEDS

Joseph Truini

The Taunton Press

The Taunton Press
Inspiration for hands-on living®

The Taunton Press, Inc.
63 South Main Street
PO Box 5506, Newtown, CT 06470-5506
Email: tp@taunton.com

Editor: Peter Chapman
Copy editor: Seth Reichgott
Indexer: Jay Kreider
Jacket/Cover design: Guido Caroti
Interior design: Guido Caroti
Layout: Guido Caroti, Sandra Mahlstedt
Illustrator: Christopher Mills
All photos by Geoffrey Gross, except as noted below:

pp. 3, 36-41, 42 (top left, top right), 43, 44 (top middle, top right), 45 (top right, bottom right), 46-63: courtesy
Fine Homebuilding magazine © The Taunton Press, Inc.
p. 8: courtesy DekBrands®
p. 9 (bottom), p.14 (bottom): © Smith-Baer
pp. 18 (left), 21 (left), 22 (left), 23 (top right), 26 (top right): courtesy Pine Harbor Wood Products
p. 26 (bottom): courtesy DaVinci Roofscapes
pp. 29 (bottom right), 44 (top left), 45 (top left, top middle), 89 (bottom right), 105 (top left, top right),
181 (bottom), 182, 184 (left), 192-93, 196 (right), 199 (right), 201 (right), 202, 206 (all, except bottom left),
208, 209 (top left, top right, bottom right): Joseph Truini

The following names/manufacturers appearing in Building Sheds are trademarks:
Azek®, CertainTeed®, Cor-A-Vent®, DaVinci Roofscapes®, DekBrands®, GAF®,Owens Corning®, Palram®,
Sonotubes®, Stanley®, SunTuf®, TimberLOK®, TimberSIL®, T-Rex™, Tuftex®, Tytan®

Library of Congress Cataloging-in-Publication Data

Names: Truini, Joseph, author.
Title: Building sheds / author, Joseph Truini.
Description: Newtown, CT : Taunton Press, Inc., 2016. | Includes index.
Identifiers: LCCN 2015045971 | ISBN 9781627107709
Subjects: LCSH: Sheds--Design and construction--Amateurs' manuals. |
    Toolsheds--Design and construction--Amateurs' manuals.
Classification: LCC TH4962 .T783 2016 | DDC 690/.8922--dc23
LC record available at http://lccn.loc.gov/2015045971

Printed in the United States of America
10 9 8 7 6 5 4 3

Construction is inherently dangerous. Using hand or power tools improperly or ignoring safety practices can lead to permanent injury or even death. Don't try to perform operations you learn about here (or elsewhere) unless you're certain they are safe for you. If something about an operation doesn't feel right, don't do it. Look for another way. We want you to enjoy working on your home, so please keep safety foremost in your mind.

To Marla, Kate, and Chris: Thanks for shining your light on me.

Acknowledgments

Thank you first of all to all the talented men and women at The Taunton Press. I'm proud to have my name on one of your books. Thanks in particular to Executive Editor Peter Chapman: This book wouldn't have been possible without your unwavering support and constant encouragement.

And my sincerest appreciation to art director Rosalind Loeb Wanke, designer Guido Caroti, and layout coordinator Sandra Mahlstedt for creating such an attractive, thoughtful book design. Thanks, too, to production manager Lynne Phillips and administrative assistant Sharon Zagata. Your contributions made my job much easier.

As the author, I sweated over each and every word, but truth be told, it's the photographs that make this book special. It was my pleasure once again to work with photographer extraordinaire Geoffrey Gross, whose keen eye and skillful approach produced the most information-rich images imaginable.

It isn't possible to write a shed book as comprehensive as this one without a lot of help. I'd like to thank Justin Fink and Rodney Diaz of *Fine Homebuilding* magazine for their assistance with the Timber-Frame Garden Shed. Many thanks to Peter Charest of Connecticut Post & Beam for allowing us to photograph the construction of the Board-and-Batten Shed and Post-and-Beam Barn. Pete and his crew, including Dennis Royer and Greg Butkus, couldn't have been more accommodating and helpful.

Thank you to contractor Doug Foulke, and his son Devon, who built the Vinyl-Sided Storage Shed. They both showed great patience and good humor over many days of construction and photography. And many thanks to Jamie McGrath of Pine Harbor Wood Products for inviting us to join his crew—Mike Anderson, Jesse Chase, and Jason Phillips—to build the Cedar-Shingle Shed, one of the most beautiful sheds to appear in any book.

Finally, I'd like to express my sincerest thanks to the following people who went out of their way to provide me with information, photographs, technical data, and products: Kathy Ziprik (DaVinci Roofscapes®), Penny Barrows (Better Barns), and Angelika Igoe and Chris Apolito (Pine Harbor Wood Products).

CONTENTS

Introduction

I've been writing about home-improvement projects for more than 30 years. And although remodeling trends come and go, and then come again, there has always been one constant topic of interest: storage sheds. Why? Because no one has ever said, "We love our home, but there's just way too much storage space."

For 20 consecutive years I wrote at least one shed-building article per year for magazines such as *Popular Mechanics*, *Handy*, and *Today's Homeowner*. And in each year, those articles were highly ranked as the most popular amongst the readers. Now I would like to believe that such high praise was due to my wonderful prose, but in truth it was because of an undeniable fact: Every homeowner needs a storage shed. Even homeowners who already have a shed are often interested in building a new one because their existing shed is too small or dilapidated and ready to collapse.

So when the fine folks at The Taunton Press asked me to write this book, I called upon all my shed-building skills and experiences to provide you with the inspiration and information to build your own backyard storage shed. And unlike most other shed books, which show lots of pretty pictures of sheds but don't actually show how to build a shed, I took a decidedly more hands-on approach.

The first chapter takes an in-depth look at all the various construction methods and building materials you'll need to build a shed. Then, in each of the next five chapters, I'll show how to build a shed from scratch. And these aren't your typical plywood-box shacks. These are beautiful, perfectly proportioned structures that would be an asset—not an eyesore—to any yard in any neighborhood.

They include an 8-ft. by 10-ft. Timber-Frame Garden Shed that combines a potting shed with a mini-greenhouse; a 10-ft. by 10-ft. Board-and-Batten Shed that has a realistic-looking faux-slate roof; a 10-ft. by 16-ft. Vinyl-Sided Storage Shed that features a maintenance-free exterior; a beautiful 12-ft. by 16-ft. Cedar-Shingle Shed, which has wood shingle walls and roof; and a massive 14-ft. by 20-ft. Post-and-Beam Barn that resembles a traditional timber-frame building but was built using time-saving modern construction methods.

For each shed, there are dozens of step-by-step photos, detailed technical drawings, and helpful hints to guide you through each phase of construction. I hope you enjoying reading this book as much as I did writing it. Happy building!

1

SHED-BUILDING METHODS AND MATERIALS

In this opening chapter, we'll discuss the wide variety of construction methods and building materials you'll need to consider before constructing a backyard storage shed. The construction method you choose will depend on several factors, including your carpentry skill level, available time, the size and complexity of the shed, and, of course, cost. The construction method affects every aspect from foundation to roofing, so taking the time now to decide on these important issues will make the project go much more smoothly and greatly enhance the overall look and life of the shed.

With regard to building materials, there's never been a better time to construct a shed. Lumberyards and home centers are fully stocked with everything you'll need, from pressure-treated wood and framing lumber to rust-resistant hardware and PVC trim boards. Here, we'll focus on four key components: siding, roofing, windows, and doors.

In each of the next five chapters we'll show how to build a specific shed from scratch. However, with the information found in this chapter you can design and build your own shed, or alter the design of one of the sheds shown in this book. Now, let's get started!

Create an on-grade foundation with parallel rows of solid-concrete blocks. Use a long, straight 2×4 and 4-ft. level to ensure that each row is level.

Shed-Building Methods

The methods used to build a shed can be divided into four construction phases: foundation, floor framing, wall framing, and roof framing. For each construction phase there are several options, but you might not be able to choose a specific construction method. That's because the local building department often dictates the type of foundation or method of framing the walls or roof. The building inspector will also approve the most-appropriate spot on your property to build the shed.

ON-GRADE FOUNDATIONS

There are two basic types of shed foundation: on-grade and frost-proof. (Again, the building inspector will want to weigh in on this decision, so be sure to check with the building department before breaking ground.)

On-grade foundations, also called "floating foundations," sit right on the ground and are sufficient for all but the very largest sheds. They're also the quickest and simplest to build because you don't have to dig deep holes or pour concrete.

If a foundation block is a little too low, shim with dense, rot-resistant material, such as pressure-treated wood, composite lumber, or asphalt roof shingles.

On-grade foundations are usually made from solid-concrete blocks or pressure-treated lumber. The concrete blocks are laid out in straight, evenly spaced rows. The floor frame is then built on top of the blocks. If the ground slopes slightly, build up the low end by stacking solid-concrete blocks or adding shims cut from weatherproof material, such as pressure-treated wood, composite lumber, or asphalt shingles. And be sure to only use solid-concrete blocks, not hollow wall blocks, which would eventually crack and crumble.

7

Precast-concrete pier blocks, laid out in parallel rows, provide a quick and easy way to support the shed's pressure-treated floor frame.

Each pyramid-shaped concrete pier block has slots molded into its top surface, which are sized to accept a 2× floor joist.

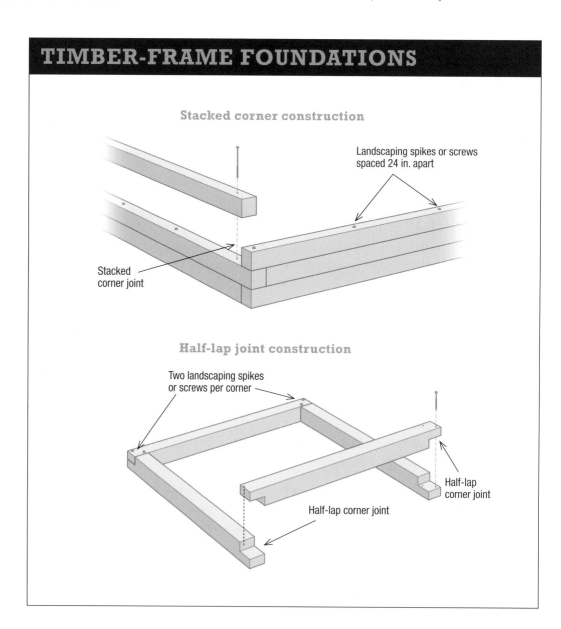

TIMBER-FRAME FOUNDATIONS

Stacked corner construction

Landscaping spikes or screws spaced 24 in. apart

Stacked corner joint

Half-lap joint construction

Two landscaping spikes or screws per corner

Half-lap corner joint

Half-lap corner joint

This skid foundation is built with three timbers, or skids, set on grade. The 2×6 floor joists are then laid across the skids spaced 16 in. on center.

Skids can be cut from solid timbers, such as 6×6s or 8×8s, or you can make them by fastening together three long 2×6s or 2×8s.

Precast-concrete pier blocks offer an alternative to solid-concrete blocks. These pyramid-shaped blocks are designed for building decks, but work great for shed foundations, too. Each block is about 8 in. high by 11 in. sq. Molded into the top surface are two crisscrossing 1½-in.-wide slots and a 3½-in.-sq. recessed socket. The slots accept a 2× floor joist; the socket is sized to hold a 4×4 post.

As mentioned earlier, on-grade foundations can also be built out of pressure-treated lumber. The two most popular methods include skid foundations and timber-frame foundations.

A skid foundation is simply made by setting two or more long, straight timbers (skids) on the ground parallel to one another. The building's floor frame is then laid across the skids. Skids are often made from pressure-treated 4×6s, 6×6s, or 8×8s, but you can also make them by gang-nailing together three or four 2×6s or 2×8s. Skids can be set directly on the ground, but it's better to first put down 2 in. to 4 in. of compacted gravel. The stone creates a more stable base and prevents erosion and settling.

A timber-frame foundation consists of a rectangular wooden frame placed on top of a compacted gravel bed. The shed walls are then fastened to the frame. The benefit of this type of foundation is that you can choose from a variety of flooring options: crushed stone, poured concrete, composite decking, bluestone slabs, or brick pavers.

The timber frame is typically made from pressure-treated 4×4s, 4×6s, or 6×6s. The timbers are joined with half-lap corner joints or are simply stacked two or three high and fastened together with long structural screws or landscaping spikes.

A timber-frame foundation rests on a compacted-gravel bed. Use a 4-ft. level to confirm that all four sides of the frame are perfectly level.

Fill the space within the timber-frame foundation with 4 in. of compacted gravel, then create a beautiful, durable floor with red-brick pavers.

FROST-PROOF FOUNDATIONS

Frost-proof foundations consist of concrete piers or wooden posts that extend down to the frost line. These types of foundations prevent freeze/thaw cycles from upsetting the shed. They're generally required in cold-weather regions for buildings larger than 200 sq. ft. However, building codes differ from state to state, so check with your local building department. Here's a quick look at the three most common types of frost-proof foundations: concrete pier, concrete slab, and pole barn.

A poured-concrete pier is simply a column of concrete poured into a hole that's dug down to the frost line. Two or more rows of piers are used to support the shed's floor frame. The floor frame is then fastened to the piers with metal bolts, post anchors, or tie-down straps. We used poured-concrete piers to support the Cedar-Shingle Shed shown on p. 142.

A poured-concrete slab is the best type of foundation for large outbuildings, especially ones that'll be used to store heavy equipment, such as woodworking machines, tractors, boats, cars, motorcycles, and snowmobiles. There are two methods of pouring a concrete foun-

Precast Piers: Dig-and-Drop Foundation

Mixing and pouring concrete piers—one at a time—is time-consuming and labor intensive. But now you can install concrete piers in a matter of minutes without breaking a sweat. How? By using precast piers. These ready-to-install piers provide an incredibly quick and easy way to build a frost-proof foundation. Simply dig a hole and drop in the pier.

Precast-concrete piers come in various sizes, and each one has an extra-wide base that creates a super-stable footing. The only drawback is that the piers are far too heavy to move by hand, so you must hire an excavator to lower the piers into the holes. Once the piers are set into place, check them for level, and then backfill around each one with soil.

A. Ready-to-install precast-concrete piers consist of an extra-wide base (footing) and tapered column. Lower the piers into the holes with a backhoe.

B. Confirm that the top of each concrete pier is level in two directions: front to back, and side to side. The piers must also be level with each other.

C. Backfill around each precast pier with soil. The piers should protrude above grade by at least 4 in. to allow air to flow beneath the shed.

Make a concrete pier by digging a hole down to the frost line, then inserting a fiber-form tube. Pour concrete into the tube to create the pier.

This pole-barn foundation has pressure-treated posts set into holes dug down to the frost line. The shed walls are then secured to the posts.

dation, but only one qualifies as frost-proof. Called a monolithic slab, the floor and perimeter foundation walls are all poured at the same time. The walls extend down to the frost line and are usually 12 in. to 16 in. thick. The shed floor is 4 in. to 6 in. thick and reinforced with wire mesh or metal reinforcing bars.

The second type of concrete foundation, known as a floating slab or an on-grade slab, is simply a 4-in.- to 6-in.-thick concrete slab that sits directly on the ground. A floating slab should never be used when the local building code calls for a frost-proof foundation.

A pole-barn foundation begins with a series of holes dug down to the frost line around the perimeter of the building. Concrete footings are poured into the bottom of each hole. Once the concrete cures, pressure-treated round poles or square timbers are place into the holes. Then, horizontal beams are bolted to the poles, creating a structural frame for the walls and roof.

A shed's floor frame is cut from pressure-treated lumber and consists of mudsills, rim joists, and floor joists that span the width of the floor.

FRAMING FLOORS

Most sheds floors are conventionally framed out of pressure-treated lumber, which is resistant to decay and wood-boring bugs. The floor frame consists of mudsills, perimeter rim joists, and floor joists.

The mudsills are the lowest wood member of any building and they sit right on the foundation. The rim joists frame the perimeter of the floor, and the floor joists sit on top of the mudsills and span the width of the building. The size and spacing of the floor joists depend on the size of the shed, but they're typically cut from 2×6s or 2×8s and spaced 16 in. on center.

Once the floor frame is built, it's usually covered with ¾-in.-thick exterior-grade plywood. Be sure to align each end-butt joint over the center of a joist, then fasten the sheets with 1⅝-in. decking screws or 2-in. ring-shank nails spaced 8 in. to 10 in. apart.

Cover the floor frame with exterior-grade plywood or oriented-strand board. Be sure all end-butt joints are centered over a floor joist.

For a more traditional-looking shed floor, install solid-wood planking. This type of flooring consists of hefty tongue-and-groove pine planks that measure 1½ in. to 2 in. thick and 6 in. to 8 in. wide. Solid-wood planking is considerably more expensive than plywood and takes longer to install, but it's much more attractive and creates a stable, rock-solid floor that won't bounce or bend. Plus, the planks can be blind-nailed through the tongue edges, so that no fasteners will be visible on the surface.

Build the wall frame flat on the shed floor and then tip it up into place. This is much easier than trying to nail up the parts one board at a time.

Post-and-beam walls are built using large posts, purlins, and beams. Here, a 4×6 header is being installed to create a window rough opening.

Traditional timber-frame construction uses mortise-and-tenon joints. The protruding tenon fits snugly into the recessed mortise and is then secured with wood pegs.

BUILDING WALLS

There are three basic ways to frame shed walls: stick-built, post-and-beam, and pole barn. And while an overwhelming majority of storage sheds are stick-built out of 2×4s, post-and-beam construction and pole-barn buildings are growing in popularity.

Stick-built construction is preferred by DIYers because it's the fastest, easiest, and most affordable way to frame walls. The term *stick-built* refers to the individual "sticks" of lumber—normally 2×4s or 2×6s— used to frame the walls. A typical shed wall consists of vertical studs, horizontal top and bottom plates, and headers above all window and door openings. Wall studs are usually spaced 16 in. or 24 in. on center.

To make building walls easier, assemble each wall frame flat on the floor deck. Then nail on the plywood sheathing or siding and tip the wall up into place.

Post-and-beam construction uses large vertical posts and horizontal beams to form the skeletal frame of the walls. There are far fewer parts in a post-and-beam wall than in a stick-built wall, but the parts are much larger and heavier. The posts and beams are usually cut from 4×6s, 6×6s, or 8×8s.

The traditional method of connecting the posts and beams is with mortise-and-tenon joinery. Each joint consists of a recessed slot (mortise) and a protruding tab (tenon). Mortises are cut by first drilling out the waste wood and then squaring up the hole with a mallet and chisel. Tenons are typically cut with a portable circular saw and handsaw. The tenon fits tightly into the mortise and is then secured with wood pegs or screws.

We employed a more-modern method to build the Post-and-Beam Barn shown on p. 172. Instead of painstakingly cutting mortise-and-

Metal T-Rex™ connectors greatly simplify post-and-beam construction. The protruding flange fits into a slot cut in the end of the beam.

The beam is securely locked in place with two ½-in.-dia. metal pins, which are driven into holes bored through the beam and connector.

tenons joints, we fastened together the wall-frame timbers with metal connectors. Each T-shaped aluminum connector is screwed in place, and then its protruding flange is slipped into a slot cut in the end of the mating timber. The joint is secured by two ½-in.-dia. aluminum pins driven into holes bored through the timbers and connector.

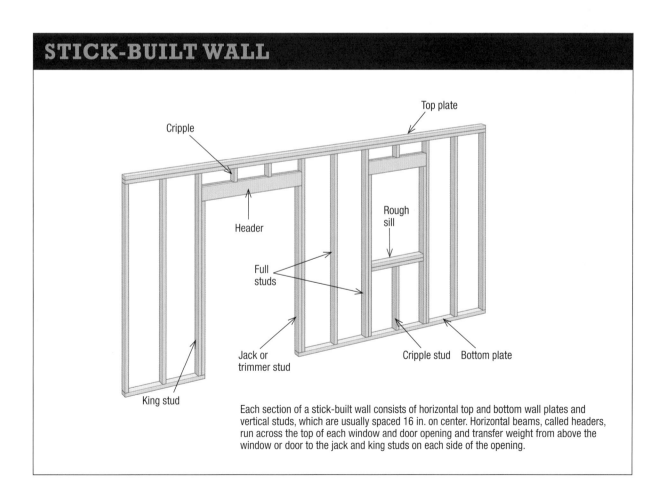

STICK-BUILT WALL

Each section of a stick-built wall consists of horizontal top and bottom wall plates and vertical studs, which are usually spaced 16 in. on center. Horizontal beams, called headers, run across the top of each window and door opening and transfer weight from above the window or door to the jack and king studs on each side of the opening.

The easiest way to frame a roof is to build roof trusses. Each truss has two rafters and a bottom chord. The parts are joined with plywood gusset plates.

In **pole-barn construction,** how you frame the walls depends largely on whether the barn has round poles or square posts. For round poles, the easiest approach is to simply nail 2×4 or 2×6 purlins horizontally across the poles. Space the purlins no more than 24 in. on center, then nail on the wall sheathing or siding.

If the pole barn is built with square posts, there are two ways to frame the walls: Nail 2×4 or 2×6 purlins across the posts, as described above for round poles, or build 2×4 wall frames and set them in between the posts. Position the 2×4 frames flush with the outer surface of the posts to simplify the installation of the sheathing and siding.

FRAMING ROOFS

The most challenging part of building any shed, especially for DIYers, is framing the roof. But roof framing isn't all that difficult, once you understand how the parts fit together.

The traditional way to frame a roof is to cut and install the rafters, ridge board, and ceiling joists individually. This piecemeal approach allows a single person to frame the roof one board at a time. The drawback is that you spend a lot of time climbing up and down ladders.

A second technique is to preassemble the roof-frame parts into roof trusses, and then raise each truss into place. This method is easier and safer than working from a ladder, but you'll need two or three people to lift the trusses onto the walls. We built trusses to frame the roof of the Board-and-Batten Shed (p. 66) and the Vinyl-Sided Storage Shed (p. 108).

Now let's take a look at five roof styles: gable, saltbox, shed, gambrel, and hip. Gable and shed roofs are by far the simplest to frame. Saltbox

Carry the preassembled roof trusses into the shed. Then, lift them up and set them on top of the sidewalls, directly over the wall studs.

This attractive 12-ft. by 16-ft. storage shed features a gable roof with a 10-in.-12 slope. Note how the roof overhangs the gable ends of the building.

and gambrel roofs are only slightly more difficult to build but are more architecturally interesting. A hip roof has four roof planes, which is distinctive looking but trickier to frame than other types of roofs.

The familiar A-shaped profile of a **gable roof** is formed by pairs of common rafters that extend from the roof peak down to the tops of the walls. A ridge board, if used, runs horizontally between each pair of rafters along the peak. Most gable roofs for sheds are framed with 2×4 or 2×6 rafters and ceiling joists, spaced 16 in. or 24 in. on center. Larger sheds would require rafters cut from 2×8s or larger lumber. Check with the building inspector for the proper lumber size and spacing. And note that roof slopes for gable sheds typically range between 6-in.-12 and 12-in.-12.

HELPFUL HINT

Set each roof truss directly over a wall stud to help transfer the weight of the roof to the wall frame and down through the floor frame to the foundation.

The gable roof on this Colonial-style shed is easily recognized by its A-shaped profile. Gable roofs are by far the most popular type of shed roof.

A saltbox roof adds a bit of Colonial charm to this 12-ft. by 14-ft. storage shed. The rear roof plane is slightly longer than the front roof plane.

This cleverly designed saltbox shed has a rear wall that's recessed 2 ft. beneath the roof overhang to create a readily accessible storage nook.

The charming **saltbox roof** is similar to a standard gable roof except that one roof plane is slightly longer than the other. This single alteration shifts the roof peak closer to the front wall, thus creating the distinctive saltbox design.

The trick to producing a true Colonial-style saltbox is to follow two strict roof-framing rules: First, cut the rafters for a 12-in-12 slope (that's 45°), and locate the roof peak one-third of the way back from the front wall. Violating either of these rules will throw off the proportions and make the entire building look distorted and out of balance.

A **shed roof** is the simplest of all roofs, consisting of a single sloping roof plane. This type of roof can be installed on freestanding outbuildings but is much more common on lean-to sheds that are built up against a house, barn, garage, or other structure. Shed roofs are generally built at a relatively shallow slope, ranging between 4-in-12 and 6-in-12.

The barn-style **gambrel roof** is easily recognized by its distinctive double-sloping profile. Gambrels are more difficult to frame than gable, shed, or saltbox roofs, mainly because they contain more parts. However, the advantage is that a gambrel is very spacious inside with lots of headroom for storing tall items or installing an overhead storage loft.

The one small drawback is that you typically have to install the doors on the shed's end wall because the sidewalls of a gambrel are too short. And like most other roof styles, a gambrel roof can be framed one board at a time or preassembled into roof trusses and then lifted on top of the walls.

A shed roof has a single sloping roof plane, which makes it ideal for lean-to sheds that are built up against another structure.

Greatly simplify the construction of a gambrel roof by building roof trusses on the ground. Then lift the trusses into place on top of the walls.

A **hip roof** is essentially a gable roof with four sloping planes. And the roof planes extend past the walls, creating an overhanging eave that wraps around the entire building, a feature that's unique to hip roofs. Hip roofs can be built onto virtually any shed, but are particularly well suited for square, hexagonal, and octagonal outbuildings where the four roof planes come to a point at the peak.

This diminutive garden shed features a distinctive hip roof, which is easily recognized by its four sloping roof planes.

This hexagonal outbuilding is topped with a hip roof that extends past each of the shed's six walls, creating an overhanging eave.

Cedar bevel siding, also called clapboards, is nailed to the walls in overlapping courses. The siding shown here comes preprimed and ready for paint.

Hip roofs are distinctive, but still not very popular, primarily because they're more difficult and time-consuming to frame than other roof types. That's partially because a hip roof has three kinds of rafters: common, hip, and jack. And it can be challenging to cut the necessary miter, bevel, and compound miter joints. Plus, hip roofs provide very little overhead storage space.

Shed-Building Materials

The first half of this chapter detailed the various construction methods used to build a shed. Now let's examine four shed-building materials that affect the overall look and style of a shed, including siding, roofing, windows, and doors.

The building materials you choose within each of these four categories will often be based on personal preference and cost, but there are practical considerations as well. For example, PVC trim costs four to five times more than wood trim, but it won't ever rot or split, and it never needs painting. Textured-plywood siding is an inexpensive alternative to most other types of siding, and it can be nailed directly to the wall framing, saving you the cost and trouble of installing plywood sheathing.

Carefully weigh all your options and you're sure to find the most-appropriate building materials for your shed and budget.

HELPFUL HINT

Save time and reduce costly measuring errors by cutting a pair of rafters and checking to make sure they're the correct length and come together to form the proper roof slope. Once satisfied, use one of the rafters as a template to mark the remaining rafters.

TOOL TIP

A power miter saw won't replace your portable circular saw, but it'll offer a quicker, more accurate way to crosscut framing lumber, especially roof rafters.

This stately little saltbox shed has cedar bevel siding, which was painted. Cedar siding is also an excellent choice if you plan to stain the shed.

Tongue-and-groove boards are widely used as vertical siding for sheds and barns. They are commonly available in pine, red cedar, and redwood.

SIDING

You can install virtually any type of house siding on a storage shed. Here, we'll discuss seven viable options. The type of siding you choose will depend on several factors, including availability, cost, ease of installation, and the style you wish to achieve: rustic or refined, casual or classic.

Cedar bevel siding, also known as clapboards or lap siding, comes in long, thin planks. It's called bevel siding because the siding is cut with the sawblade tilted at a slight bevel angle. As a result, the planks are thinner along the upper edge and thicker at the bottom, or butt, edge. When installed, the upper edge of each siding course is overlapped by the thicker butt edge of the course above it.

Note that the bevel siding is rough-sawn on one side and smooth-sanded on the other. If you plan to stain the siding, install it with the rough side facing out. The rough-textured surface will readily absorb the stain. If you'll be painting the siding, put the smooth side out. We installed cedar bevel siding on the lower half of the walls on the Timber-Frame Garden Shed shown on p. 30.

One of the most popular types of vertical-board siding is V-jointed **tongue-and-groove siding.** We installed pine tongue-and-groove siding on the Post-and-Beam Barn shown on p. 172. This type of siding is also available in western red cedar and redwood, which cost more than pine, but these two softwood species are naturally resistant to splitting, decay, and voracious termites.

HELPFUL HINT

If you live in an area that receives significant snowfall, the shed's roof must satisfy a specific snow-load rating. To find out if your shed roof meets the building-code requirement, check with the building inspector.

A quicker way to create board-and-batten siding is to sheathe the walls with rough-sawn plywood and then nail on equally spaced wood battens.

Shingles are fastened to the walls in overlapping courses. To ensure a weatherproof surface, all vertical seams must overlap by at least 1 ½ in.

This spacious seaside shed features two types of siding: white cedar shingles on the gable-end walls and board-and-batten siding on the sidewalls.

The ¾-in.-thick boards are milled with a tongue along one edge and a groove along the other. When installed, the tongue of one board fits tightly into the groove of the adjacent board. And ordinarily the edges are chamfered to 45°, creating a decorative V-shaped joint along the vertical seams.

Tongue-and-groove siding can be nailed directly to the wall framing; there's no need to first sheathe the walls with plywood. However, when installing the siding vertically you must install horizontal blocking or purlins to the wall framing to provide a solid nailing surface.

Board-and-batten siding is another very popular style of vertical-board siding. It consists of wide boards and narrow wood strips, called battens. The boards are nailed vertically to the wall frame, and then the battens are fastened over the seams between the boards.

The boards can be made from wood planks ranging in width from 1×6s all the way up to 1×12s. The battens are typically cut from 1×3s. And you can mix and match the boards to create variable-width board-and-batten siding. When installing the siding, leave a 1-in. gap between the boards, so they can expand without buckling. And fasten the battens over the gaps, making sure they overlap the boards by a minimum of ½ in. That way, the gaps will remain concealed, even if the boards shrink.

We installed board-and-batten siding to the upper half of the walls on the Timber-Frame Garden Shed shown on p. 30. For the Board-and-Batten Shed shown on p. 66, we devised a shortcut and applied battens to textured plywood, thus eliminating the individual boards.

This beautiful pool cabana is festooned completely in wood shingles: red-cedar shingles on the roof and white-cedar shingles on the walls.

Textured plywood siding comes in large sheets for quick, easy installation. And it can be nailed directly to the wall framing; no sheathing needed.

To create the look of a seaside cottage, consider installing **cedar shingles.** Cut from eastern white cedar, the individual shingles are installed in overlapping courses. The nails in each course are covered by the course above, resulting in a neat, clean appearance.

Cedar shingles aren't quite as popular as other types of wood siding because they're time-consuming to install and relatively expensive. And you must first sheathe the walls in plywood before nailing up the shingles.

If you'd like to create a more rustic-looking exterior, consider hand-split cedar shakes. Shakes are similar to shingles, except that they're much thicker and more heavily textured.

Textured-plywood siding provides the quickest, most economical way to side a shed. The $5/8$-in.-thick exterior-grade plywood comes in 4-ft.-wide by 8-ft.-long sheets, so you can cover large expanses of wall very quickly. And you can nail the plywood siding right to the studs, without first sheathing the walls.

Plywood siding comes in several styles, including rough-sawn, primed, and unprimed. The most popular style by far is grooved plywood siding, which is commonly called T-1-11. This type of siding has a rough-sawn surface that features a series of equally spaced $3/8$-in.-wide grooves. You can buy the siding with the grooves spaced either 4 in. apart or 8 in. apart. The 4-in. spacing looks best on smaller sheds. Choose the 8-in. pattern for larger outbuildings.

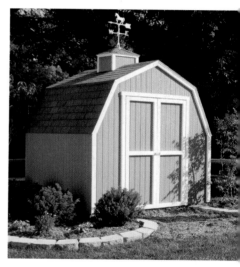

This mini-gambrel storage shed features grooved, textured plywood siding.

Fiber-cement siding is installed much like bevel siding, in overlapping courses. However, fiber-cement is extremely durable and fireproof.

Vinyl siding is lightweight, easy to cut, and a snap—literally—to install. Each siding panel snaps onto the course below for quick installation.

HELPFUL HINT

To prevent paint or stain from peeling and fading prematurely, coat the back of each piece of siding before installing it. This technique, known as back-priming, seals the boards and prevents moisture from passing through the siding from the back surface.

Fiber-cement siding is a super-durable, low-maintenance building product that looks and installs somewhat like cedar bevel siding. The $5/16$-in.-thick by $6\frac{1}{8}$-in.-wide siding has an embossed wood-grain surface that resembles real wood, but it's composed entirely of sand, cement, and cellulose fibers.

Fiber-cement siding is ideal for sheds because it won't rot or crack, and its surface accepts paint beautifully. It's also highly resistant to moisture, termites, and fire. However, just like cedar bevel siding, fiber-cement siding must be nailed to plywood wall sheathing.

Vinyl siding is, in many ways, an ideal shed-siding material: It's easy to install, never needs painting, won't ever peel or rot, is extremely weather resistant, and comes in a wide array of colors. Even so, vinyl siding is seldom used on outbuildings, and there are a few reasons why.

First, most DIYers aren't sure how to install vinyl siding, or even where to buy it. (For the record, vinyl siding is easier to install than most wood siding, and it's available by special order from most lumberyards.) Second, vinyl siding doesn't stand up very well to the inevitable bumps and bruises suffered by storage sheds. Run-ins with lawnmowers, ladders, lumber, and other items can punch holes in the siding, crack it, or knock loose a siding panel.

Despite its drawbacks, some homeowners choose vinyl siding for their sheds because they want to match the siding on their homes. And since vinyl is the most popular house siding—roughly one-third of new homes have vinyl siding—we're likely to see more vinyl-sided sheds. For specific details on installing vinyl siding, see the Vinyl-Sided Storage Shed on p. 108.

Architectural-style roof shingles are thick and heavily textured. The roof shown here is designed to look somewhat like weathered wood.

The architectural-style roof shingles on this rustic outbuilding feature the deep shadow lines and variegated greenish-blue tones of natural slate.

ROOFING

Shopping for roofing might not be as exciting as choosing siding or windows, but don't underestimate the impact a roof can have on the shed's overall look and style. Here, we'll take a look at the three most popular DIY roofing options: asphalt shingles, cedar shingles, and plastic-composite shingles.

Asphalt shingles are by far the most popular type of shed roofing, and it's easy to see why: They're affordable, easy to install, readily available in several colors, and surprisingly durable; many brands carry 40-year warranties.

There are two main types of asphalt shingles: three-tab and architectural-style. Standard three-tab shingles represent a very basic, generic style. Each shingle is a single layer thick, with two narrow slots cut into it to create the three tabs. Architectural-style shingles consist of two strips of asphalt roofing, one laid on top of the other. The bottom strip is a solid shingle and the top strip is notched with widely spaced, dovetail-shaped tabs. When installed, the laminated construction forms a heavily textured surface with deep shadow lines.

Few roofing materials can compare with the natural beauty, color, and texture of **western red-cedar shingles.** The shingles are nailed down in overlapping courses, with each succeeding course covering the nails in the previous course below.

Red cedar is naturally resistant to rot and wood-boring bugs, but offers very little resistance to fire, unless it's treated with a fire retardant. Cedar shingles are available in several grades. Use Number 1 Blue

Western red-cedar roof shingles are typically nailed down to spaced sheathing, which allows air to circulate beneath the shingles.

The natural beauty and texture of red-cedar roof shingles are showcased on this backyard shed. The shingles will eventually weather to light silvery-gray.

Label shingles for roofing, which is a premium-grade shingle that's perfectly clear (no knots) and cut from all-heartwood for superior decay resistance.

Note that cedar shakes are also available for use as roofing. Shakes are thicker and rougher than shingles, so they have a much more rugged and rustic appearance.

Plastic-composite shingles represent the newest and most intriguing type of roofing. These resilient shingles are rot-proof, weatherproof, fire-resistant, and virtually indestructible. They're often called faux shingles because they resemble natural slate tiles or cedar shingles or shakes.

Plastic-composite roofing is installed similar to other shingles: It's nailed down to plywood sheathing in overlapping courses. The time, tools, and skill required to install composite shingles are similar to asphalt roofing, but faux shingles cost three to four times more. We installed a faux-slate roof on the Board-and-Batten Shed shown on p. 66.

The beautiful red-cedar shakes on this home are not cedar at all, but plastic-composite roof shingles made by DaVinci Roofscapes.

This ruggedly handsome roof could easily be mistaken for natural slate but is actually built out of plastic-composite shingles.

Barn-sash windows are specifically made for use in sheds, barns, stables, and other outbuildings. The sash typically tilts in for ventilation.

WINDOWS

Shed windows serve two basic functions: to provide natural light and to admit fresh air. But they also add character and style to a shed. Generally speaking, any window that you'd put in a house can be installed in a shed, including double-hung, casements, sliders, and awning windows.

The size and number of windows you'll need will depend primarily on the size of the shed. Small to medium-size outbuildings look best with smaller and fewer windows. Larger sheds and barns can support larger and more windows.

Most windows have frames made of wood, aluminum, or vinyl. Clad windows have wooden frames with an exterior covering—or cladding—of aluminum or vinyl. Clad windows are popular because they combine a warm wood interior with a low-maintenance exterior that doesn't require painting.

Another good option for outbuildings is a barn-sash window, which technically isn't a window, but just a window sash. It's comprised of a glass pane set into a simple pine frame. Since barn sash don't come installed in a frame, you must build a frame into the rough window opening. We installed barn-sash windows in the Board-and-Batten Shed shown on p. 66.

Barn sash are commonly available at farm-supply stores and lumber-yards in 2-ft. by 2-ft. and 2-ft. by 3-ft. sizes. Be sure to paint or stain the sash's wood frame prior to installation.

Aluminum sliding windows don't possess the traditional look or charm of double-hung windows, but they do have a few features that make them suitable for sheds. First, the durable aluminum frames never need painting, and they come with insect screens. Plus, when the sash is opened it's contained within the frame; it doesn't swing out or

What aluminum windows might lack in architectural beauty, they make up for in durability, ease of installation, and very low maintenance.

Sliding doors glide open on two roller assemblies, called trucks. The trucks roll along a horizontal steel rail mounted to the shed wall.

lean into the shed. Aluminum windows come in dozens of sizes, but the two most popular models for outbuildings are 2 ft. by 4 ft. and 2 ft. by 6 ft.

DOORS

The type of door to install depends on the size and style of the shed, and what's stored inside. A single door is sufficient for storing household items, but a pair of double doors is necessary for housing a lawn tractor, snow blower, and other equipment.

Shed doors can either be hinged or sliding. Hinged, swinging doors are inexpensive and easy to install. And outswinging doors don't take up any floor space inside the shed.

Sliding doors glide completely out of the way, cover wider openings than hinged doors, and can't be blown shut by the wind. And because sliding doors fit over—not into—the doorway opening, you don't have to worry about making the door fit precisely. Sliding doors are usually mounted to the outside of the shed, but you can also hang them inside, as shown for the Vinyl-Sided Storage Shed on p. 108.

Sliding doors are also a good option for sheds built in areas that receive a lot of snowfall. Unlike an outswinging door, a sliding door can be opened without having to first shovel the snow out of the way.

Here are brief descriptions of how to build two different types of shed doors: batten doors and frame-and-panel doors. Neither type is particularly difficult to build, but if you don't have the time or inclination to build a door from scratch, then simply buy one. Home centers and lumberyards stock several styles of prehung exterior-grade doors made from wood, steel, and fiberglass.

HELPFUL HINTS

• The most useful shed-door design includes a single door on one wall and a pair of double doors on another. That way, you've got two points of entry and the ability to store small and large items.

• You can save a little money by building shed doors out of pine lumber, but red cedar is much more weather resistant and less likely to warp or rot.

• Be sure the lumber used to build batten doors is thoroughly dry. Otherwise, the doors will shrink when the lumber dries.

Build batten doors on a large, flat workbench. Glue and screw the battens to the rear of the door panel in a Z-shaped pattern.

When building a plywood frame-and-panel door, apply construction adhesive to the wooden frame parts, then screw them to the plywood panel.

A **batten door** consists of a panel made from several vertical boards, which are held together with wooden strips, called battens. The battens are glued and screwed to the back of the door panel in an X- or a Z-shaped pattern. You can make the panel from standard, square-edged 1×6s, but it'll be much stronger if you use tongue-and-groove, V-jointed 1×6s.

To build a batten door begin by cutting all the tongue-and-groove 1×6s to length for the door panel. Then, use a tablesaw to rip the groove off of one board and the tongue off another, creating two square-edged boards. Lay the 1×6s face down on a flat surface, starting and ending with a square-edged board. Clamp the boards together to form the door panel.

Now cut the battens from standard square-edged 1×6s. Apply construction adhesive to the battens, and then screw them to the back of the door panel. Be careful the screws don't poke through the door.

A **frame-and-panel plywood door** is one of the quickest, easiest types of shed doors to build. Start by cutting a plywood door panel to fit the doorway opening. Then cut 1×4s or 1×6s to form a.frame around the perimeter of the plywood panel. Glue and screw the frame parts to the face of the panel.

Use ⅝-in.- or ¾-in.-thick plywood for the panel, and cut the frame from cedar or pressure-treated wood. Note that you can also make a door panel from tongue-and-groove boards, as shown for the Cedar-Shingle Shed on p. 142.

The back of this batten door shows how the vertical-board door panel is held together by three horizontal battens and two diagonal battens.

2

TIMBER-FRAME GARDEN SHED

Every home gardener needs a dedicated place to store tools, organize supplies, mix soil, fill pots, and prune and propagate plants. And this distinctive 8-ft. by 10-ft. garden shed certainly satisfies all of those requirements, but that's not all. This versatile outbuilding also serves as a mini-greenhouse.

One side of its gable roof is covered with beautiful red-cedar shingles, but the opposite side is sheathed with clear polycarbonate panels that allow sunshine to fill the interior space, creating the perfect environment for growing plants and jump-starting seedlings.

And to make this shed as attractive on the inside as it is from the outside, its walls and roof are framed out of 4×4s and 4×6s, giving it the appearance of a traditional timber-frame structure.

The shed also features two types of siding: cedar bevel siding on the lower half of the walls and board-and-batten siding on the upper half. Seven double-hung windows admit plenty of sunlight and fresh air. And because this shed takes up only about 80 sq. ft. of space, it'll comfortably fit into the tiniest yards and smallest gardens. (To order a set of building plans for the Timber-Frame Garden Shed, see Resources on p. 214.)

Timber-Frame Foundation

The shed is supported by a foundation of pressure-treated 4×4 timbers that sit on two parallel rows of solid-concrete blocks. This type of on-grade foundation is quick and easy to build and doesn't require mixing and pouring concrete. Plus, there's very little excavation to do, even if the site is uneven or sloping slightly.

The perimeter of the shed's floor frame is composed of eight 4×4 timbers: two stacked along each of the four sides. There are also three 4×4 floor joists that span the width of the floor frame and sit in notches cut into the upper course of the perimeter 4×4 timbers.

CUT THE FLOOR FRAME PARTS

Start by cutting to length the 11 pressure-treated 4×4s that make up the floor frame. The best tool for quickly and accurately cutting 4×4s is a 10-in. or larger power miter saw. You could use a portable circular saw, but since it doesn't cut deep enough to saw through a 4×4 in a single pass, you'll have to make one cut, flip over the 4×4, and make a second cut.

The perimeter of the 8-ft. by 10-ft. floor frame consists of two courses of 4×4 rim joists, one stacked atop the other. The 4×4s overlap at the corners, creating strong, easy-to-assemble joints. For each of the two ends of the frame, cut one 4×4 to 89 in. for the lower course and another to 96 in. for the upper course.

For each of the two long sides of the floor frame, cut one 4×4 to 120 in. for the lower course and another to 113 in. for the upper course. Then cut three 4×4 floor joists to 92½ in. You'll now have all 11 floor-frame parts cut to length.

Next, mark the two 113-in.-long upper side rim joists for notches that will accept the three floor joists. (The 120-in.-long lower side rim joists don't get notched.) Set the upper rim joists on sawhorses and clamp them together with their ends flush. Then measure and mark the locations of the three floor-joist notches (see the drawing on p. 37).

When laying out the notches, use a short 4×4 block as a marking gauge. Hold the block in position with a layout square (1), then mark the width of each notch across both rim joists (2).

Since pressure-treated 4×4s vary in size depending how wet they are, use a 4×4 block to mark the width of the rim-joist notches.

After positioning the 4×4 block with a layout square, draw a line along each side of the block to represent the shoulder cuts.

TIMBER-FRAME GARDEN SHED

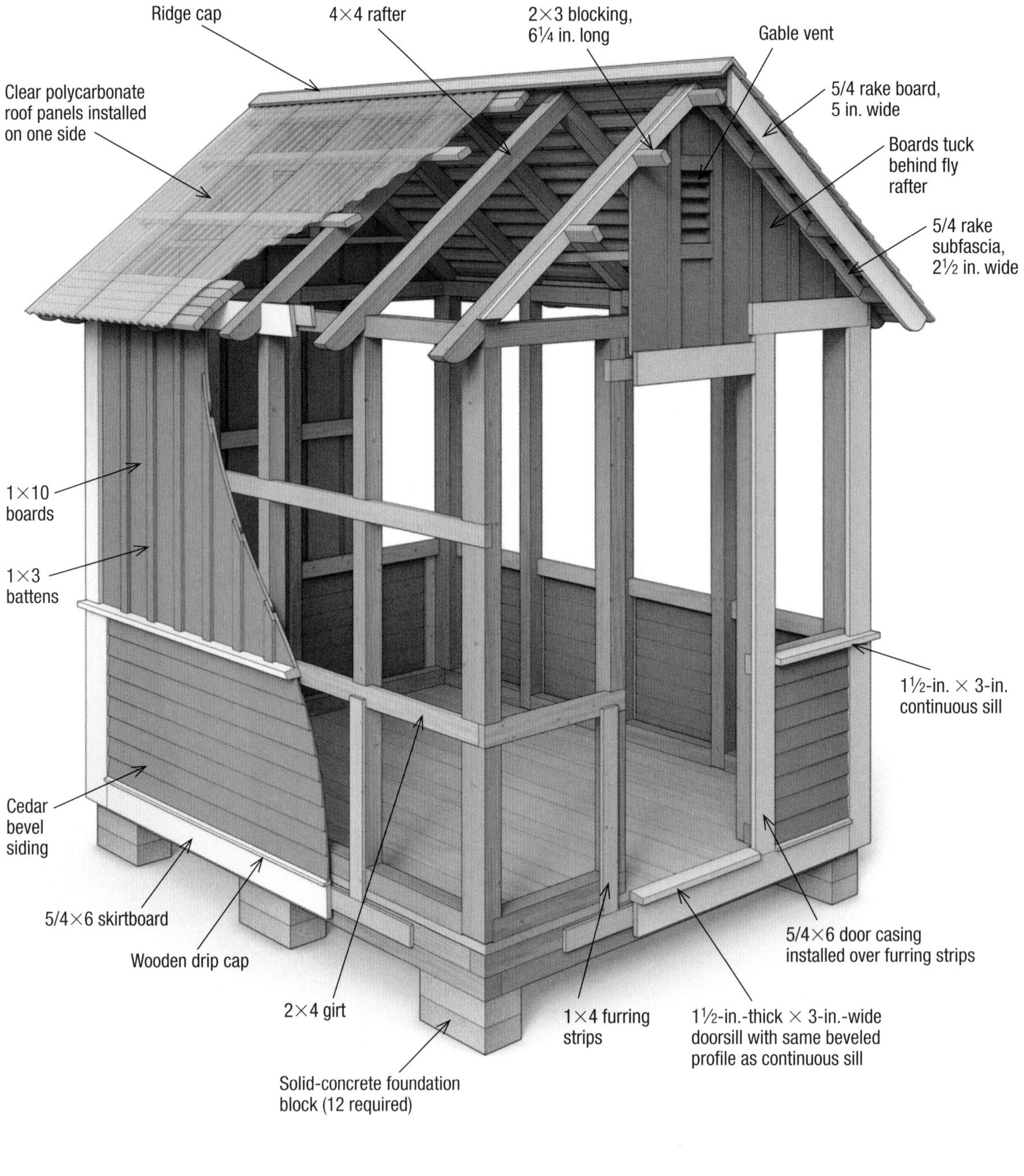

Ridge cap
4×4 rafter
2×3 blocking, 6¼ in. long
Gable vent
Clear polycarbonate roof panels installed on one side
5/4 rake board, 5 in. wide
Boards tuck behind fly rafter
5/4 rake subfascia, 2½ in. wide
1×10 boards
1×3 battens
1½-in. × 3-in. continuous sill
Cedar bevel siding
5/4×6 skirtboard
Wooden drip cap
5/4×6 door casing installed over furring strips
2×4 girt
1×4 furring strips
1½-in.-thick × 3-in.-wide doorsill with same beveled profile as continuous sill
Solid-concrete foundation block (12 required)

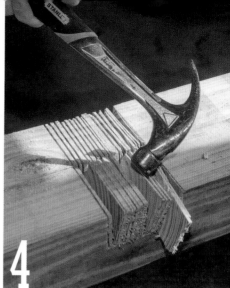

3 Guide the circular saw along a layout square to produce accurate, perfectly square shoulder cuts on each side of the half-lap joints. Then saw away a majority of the waste wood in between by making several closely spaced kerf cuts.

4 Once you've made all the kerf cuts, use a hammer to break off the thin waste-wood slices that remain in the rim-joist notches.

Next, take a circular saw and adjust its depth-of-cut to 1¾ in. deep, which is half the thickness of a 4×4. Then, carefully cut along the marked lines of each notch to establish the shoulders of the notches (**3**). Now make a series of closely spaced kerf cuts through the waste area between each pair of shoulder cuts. Space the cuts no more than ¼ in. apart.

After kerf-cutting all three notches across both rim joists, use a hammer to break away as much wood as possible from the notches (**4**). Then use a wide wood chisel to scrape the bottom of the notches until they're smooth, flat, and free of any high spots (**5**).

BUILD THE FLOOR FRAME

The first step to building the floor frame is laying out the shed's concrete-block foundation. The floor frame is supported by 12 solid-concrete blocks, which are stacked two high and arranged in two parallel rows. Each row has three two-stack blocks, and each block measures 4 in. thick by 8 in. wide by 16 in. long. If your building site slopes slightly, you'll need to build up the low end with additional 4-in. solid-concrete blocks or 2-in.-thick concrete patio blocks.

The two rows of foundation blocks are spaced 8 ft. by 10 ft. apart, as measured to the outside edges. And each of the six stacks of blocks sits on a bed of gravel. So, start by setting the blocks in position on the ground in two parallel rows. To confirm that the blocks are square to one another, measure the diagonal distance from one corner block to the next corner block. Then measure the opposite corner-to-corner diagonal distance.

5 Take a wide chisel and scrape the notches smooth and flat. Hold the chisel bevel-side down to prevent it from gouging the surface.

Starting at the highest corner, clamp a level to the 4×4 rim joist, then adjust the other blocks in the row until they're perfectly level.

Check the frame for square by measuring diagonally across the opposite corners. The frame is square when the dimensions match.

Use a cordless impact driver to quickly drive in the 6-in.-long structural screws. Secure each overlapping corner joint with two screws.

Snap a chalk line, then use a circular saw to trim the overhanging ends of the 2×6 and 2×10 floorboards flush with the shed's floor frame.

If the two dimensions are the same, the blocks are laid out square. If not, shift the blocks in or out until the dimensions are identical.

Now use line-marking spray paint, or white flour sprinkled from a can, to mark a rectangular outline around each block. Make the outlines approximately 6 in. wider and 6 in. longer than the concrete blocks. Move the blocks out of the way and dig out the dirt within each outline down to a depth of at least 6 in.; go down 12 in. for low-lying or soggy sites.

Fill each hole with 2 in. to 4 in. of gravel, then thoroughly compact the gravel using a hand tamper or 8-ft.-long 4×4. Add another 2 in. to 4 in. of gravel and tamp again. Continue in this manner until the gravel bed is flush with the surrounding surface. Note that thorough compaction is important because it combats erosion and prevents the gravel—and foundation blocks—from sinking into the soil.

Once all six gravel beds have been compacted, set the solid-concrete blocks back into place, making sure the two rows are spaced 8 ft. wide by 10 ft. long. Then, measure the two opposite diagonal distances once more to confirm that the blocks are laid out square.

Next, set a 120-in.-long lower side rim joist on top of the blocks. Place a level on top of the 4×4 rim joist and raise or lower it until it's perfectly level (**1**). If necessary, remove some gravel from beneath the block, or build it up with solid-concrete blocks, patio blocks, or 16-in.-long pieces of pressure-treated or composite decking.

Set into place the opposite 120-in.-long lower side rim joist, followed by the two 89-in.-long lower end rim joists. Then install the four upper rim joists, making sure to overlap the 4×4s at each corner of the floor frame (see the Floor Framing drawing on the facing page).

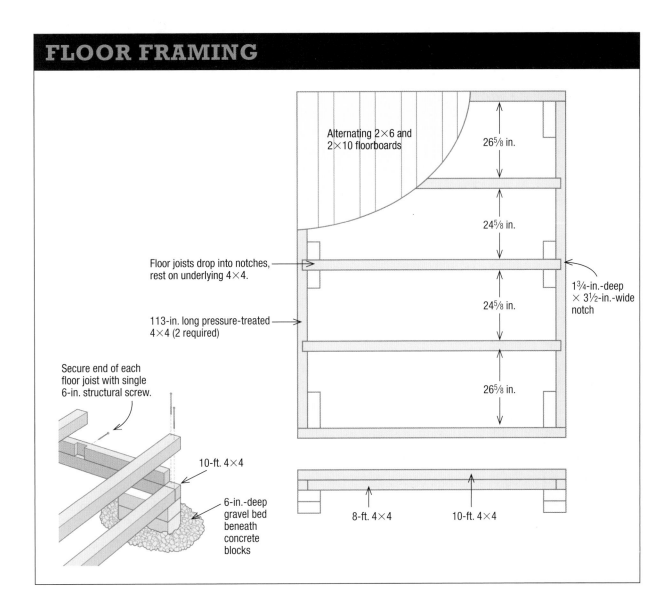

Alternating 2×6 and 2×10 floorboards

26⅝ in.

24⅝ in.

Floor joists drop into notches, rest on underlying 4×4.

24⅝ in.

1¾-in.-deep × 3½-in.-wide notch

113-in. long pressure-treated 4×4 (2 required)

26⅝ in.

Secure end of each floor joist with single 6-in. structural screw.

10-ft. 4×4

6-in.-deep gravel bed beneath concrete blocks

8-ft. 4×4   10-ft. 4×4

Fasten the upper course of 4×4 rim joists to the lower course with 6-in.-long structural screws; space the screws about 24 in. apart. Check the floor frame for square by measuring the two corner-to-corner diagonal distances (2). If the measurements aren't the same, adjust the frame in or out until they are. Now fasten together each overlapping corner of the floor frame with two 6-in.-long structural screws (3).

Install the three 4×4 floor joists, making sure each one fits snugly into the notches cut in the upper side rim joists. Secure each floor joist with a 6-in.-long screw driven into each end.

With the floor frame completed, you can install the shed floor. Nowadays, most shed floors are made of plywood or oriented-strand board (OSB), but for this garden shed we wanted something a bit more rugged and interesting looking. Here, we used construction-grade 2×6s and

HELPFUL HINT

Use only solid-concrete blocks to support a shed's floor frame. Hollow-concrete blocks, which are meant for building walls, would eventually crack and crumble under the weight of the shed.

Fast Track: Routing Joints Smooth

Using a hammer and chisel to clean up half-lap joints works well, but it can be a bit tedious and time-consuming. A faster option is to use a router fitted with a long straight-cutting bit or mortising bit.

After kerf-cutting the joints, knock away most of the waste wood with a hammer. Then, clamp a straightedge guide to each side of the notched joints. Position the guides so the router bit cuts flush along the shoulders of the joints but doesn't cut into them.

Next, adjust the router's base to the proper depth-of-cut. Set the router down onto the 4×4s, making sure the bit isn't contacting the wood. Flip on the switch and wait for the motor to reach maximum speed. Now, slowly guide the router along one straightedge guide, then back and forth across the joints, and finally up against the opposite straightedge guide.

Repeat, if necessary, until you've flattened and smoothed the bottom surface of all the half-lap joints.

Clamp together several pressure-treated 4×4 wall posts, then kerf-cut the 1½-in.-deep notches to receive the horizontal 2×4 girts.

2×10s. The boards are arranged in an alternating, repeating pattern: 2×10, followed by a 2×6, then another 2×10, and so on.

Butt the floorboards tightly together and face-nail them to the joists with 3¼-in.-long (12d) ring-shank nails. After installing all the floorboards, use a portable circular saw to rip the last floorboard flush with the side of the floor frame. Then, trim the overhanging ends of the floorboards flush with the end of the floor frame (4) (see p. 36).

Timber-Frame Walls

As mentioned earlier, the walls of the shed are designed to resemble a traditional timber-frame structure. The top and bottom wall plates and the vertical wall posts (studs) are cut from 4×4s. The four corner posts are made from beefier 4×6s. All the parts are notched with half-lap joints to lock tightly together.

When cutting the wall-frame notches into the timbers, employ the same five-step technique used earlier to notch the floor frame:

1. Clamp the parts together.
2. Cut along the shoulder lines.
3. Kerf-cut through the joints.
4. Chop out the waste wood.
5. Scrape the joints smooth.

Another advantage of using half-lap joints is that once the parts are notched the wall frame goes up surprisingly fast.

After sawing a series of closely spaced cuts through the half-lap joints, use a hammer to chop out as much waste wood as possible.

With the 4×4 wall posts still clamped together, scrape smooth the bottom of the half-lap joints with a wide wood chisel.

CUT THE WALL PARTS

Begin by cutting to length 11 wall posts, making each 4×4 post 93 in. long. Then, cut four 4×6 corner posts to 89¾ in. long each.

Saw to length four 4×4s to 113 in. to serve as the top and bottom wall plates for the shed's two long sidewalls. For the shorter gable-end walls, cut two 4×4s to 96 in. for the top wall plates and two 4×4s to 92½ in. for the bottom plates. Also, cut to length the horizontal 2×4 girts that span the wall posts.

Now, with all the wall posts and plates trimmed to length, you can begin cutting the half-lap joints (see the Wall Framing drawing on p. 40). Start by clamping together several 4×4 wall posts. Adjust the saw's depth-of-cut to 1½ in., then kerf-cut the 3½-in.-wide notches to receive the horizontal 2×4 girts (1). Use a hammer to bust out all the waste wood (2), then scrape the joints smooth with a wood chisel (3).

Next, notch both ends of each wall post to fit into corresponding notches cut into the top and bottom wall plates. However, only the bottom ends of the four 4×6 corner posts are notched; their top ends are cut square to provide flat support for the top wall plates.

When notching the post ends, cut each half-lap joint to 1¾ in. deep by 3¼ in. wide, not 3½ in. wide. Notching the half laps ¼ in. short allows the wall posts and corner posts to accommodate any shrinkage in the top and bottom wall plates.

SAFETY FIRST

After working with pressure-treated lumber, wash your hands in warm, soapy water, especially before eating or handling food.

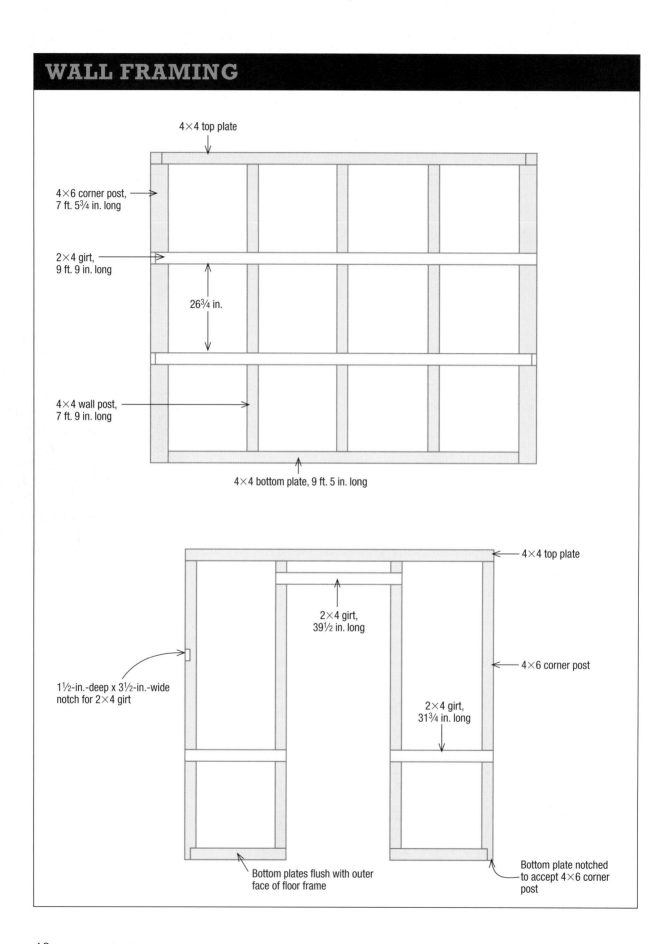

4×4 top plate

4×6 corner post,
7 ft. 5¾ in. long

2×4 girt,
9 ft. 9 in. long

26¾ in.

4×4 wall post,
7 ft. 9 in. long

4×4 bottom plate, 9 ft. 5 in. long

4×4 top plate

2×4 girt,
39½ in. long

4×6 corner post

1½-in.-deep x 3½-in.-wide
notch for 2×4 girt

2×4 girt,
31¾ in. long

Bottom plates flush with outer
face of floor frame

Bottom plate notched
to accept 4×6 corner
post

Install a pressure-treated 4×6 corner post at each corner of the shed floor. Fasten the bottom of each post with two 6-in. screws.

Cut the 2×4 horizontal girts to length, then nail them to the notches in the corner posts. Be sure the posts are plumb before nailing.

FRAME THE WALLS

To build the walls, start by fastening the 4×4 bottom plates to the shed floor with 6-in. structural screws. Position the bottom plates perfectly flush with the outside perimeter edge of the floor. Also, be sure the half-lap notches cut in the plates face in toward the center of the shed. Space the screws 24 in. to 36 in. apart.

Next, attach the four 4×6 corner posts. Secure the lower end of each post to the bottom wall plate with two 6-in. structural screws. Position the corner posts with the 1½-in.-deep girt notches facing out (1).

With all four corner posts securely screwed in place, install the lower row of horizontal 2×4 girts. Tap the girts into the notches cut in the corner posts. Check the posts for plumb in two directions, then fasten the girts with 3-in.-long nails or screws (2).

Set the 4×4 wall plates on top of the corner posts. Attach the gable-end wall plates first, then set the longer sidewall plates in between.

TOOL TIP

The best tool for driving in 6-in.-long structural screws is a cordless impact driver. Impact drivers have four main advantages over standard cordless drill/drivers: faster motor speed, much more power, less cam-out (bit slippage), and virtually no wrist-wrenching reactionary torque.

Slip the 4×4 wall posts into place, making sure their notches fit snugly over the top and bottom wall plates and horizontal 2×4 girt.

Once the 4×4 wall posts have been screwed to the top and bottom wall plates, nail through the horizontal 2×4 girt and into the posts.

Structural Screws: Reducing Lag Time

For many years, lag screws were the only viable option for fastening together large, thick timbers. While you could certainly use lags to build this shed's timber-frame, we found a better alternative: structural screws.

To drive in a lag screw (shown at left in the photo above), you must drill a counterbore for the head and washer, a screw-shank clearance hole in the top timber, and a pilot hole in the bottom timber.

Structural screws (shown at right above), on the other hand, are long and thin and can be driven in flush without any predrilling. That not only saves time but also battery power when using a cordless drill/driver.

Set the two shorter gable-end wall plates on top of the corner posts. Secure each plate by screwing down into the tops of the posts with 6-in. structural screws. Then, install the two remaining top plates in between the gable-end plates (3) (see p. 41). These longer 4×4 plates run along the top of the shed's sidewalls. Screw both plates down to the corner posts, but also run a screw through the gable-end plates and into the ends of the sidewall plates.

Now stand each 4×4 wall post into position, making sure its half-lap joints fit into the notches in the top and bottom wall plates, and over the 2×4 girt (4). Space the wall posts 24⅝ in. apart along the sidewalls and 18 in. apart in the rear wall. In the front gable-end wall, place the two wall posts 24¾ in. from the corner posts, where they'll frame the 32½-in.-wide rough opening for the door.

Fasten the lower end of each wall post to the bottom wall plate with two 2½-in.-long structural screws. Then, plumb up the posts with a 4-ft. level and screw them into the top wall plate. With the wall posts securely screwed to the top and bottom plates, nail through the 2×4 girt and into each post (5).

Now install the upper row of 2×4 horizontal girts, fastening them to the wall posts with 3-in. nails (6). Note that in order to accommodate the shed's seven 57-in.-tall windows, there are no upper girts installed in the right-hand sidewall or front gable-end wall. And in the rear wall, the upper girt doesn't extend all the way to the shed's corner. Again, that's necessary to accommodate a window.

Finish up the wall framing by nailing a short horizontal girt over the doorway opening in the front gable-end wall. This 39½-in.-long 2×4 serves as a header (7).

Install the upper row of 2×4 girts into the half-lap notches cut into the 4×4 wall posts. Secure the girts to the posts with 3-in. nails.

Nail a short girt over the doorway to create a header. The lower girt spanning the opening will be cut out later to accommodate the door.

1 Hold the rafter-layout jig flush with the end of the 4×4 rafter. Then scribe a line along the template's curved shape to mark the rafter tail.

2 Cut the curved shape into each rafter tail with a jigsaw and extra-long 6-in. blade. Cut slowly and leave just a little of the pencil line.

3 Clamp together all the rafters, making sure the rafter tails are flush. Then use a belt sander and 60-grit belt to smooth the rounded ends.

Rafter-Layout Jig

¼-in. plywood template

1¼-in. screw

4⁷⁄₈ in.

16.3°

4¾ in.

1×4

4¾-in. radius

Make the jig shown here to quickly and accurately mark the cuts onto each end of all the 4×4 roof rafters. The jig consists of two ¼-in. plywood templates fastened to a short 1×4.

The overall length of the plywood templates isn't critical, but each one must be 4¾ in. wide, with one end miter-cut to 16.3° the and other end rounded to form a 4¾-in. radius. (The templates on our jig were about 9¾ in. long.)

It's also important that the 1×4 is flush with the rounded end of the template, and extends past the mitered end of the template by precisely 4⁷⁄₈ in. Secure each template to the 1×4 with three 1¼-in. screws.

Timber-Frame Roof

The shed roof is built using the same material as the walls: pressure-treated 4×4s. However, rather than framing the roof with individual pieces of lumber, we assembled roof trusses on the ground and then lifted them up into position. You'll need help raising the five trusses onto the walls, but building and installing trusses is easier than framing the roof one 4×4 at a time.

The trusses are cut to form a 9-in-12 roof pitch, and the upper end of each 4×4 rafter is notched to create half-lap joints. A shallow bird's-mouth cut is sawn into the rafters to provide a flat bearing surface atop the walls. Also, since the rafter tails are exposed along the sidewalls of the shed, we rounded each one to make them more interesting looking.

ASSEMBLE THE ROOF TRUSSES

Start by using a power miter saw to cut 10 rafters to 6 ft. long each. At this point, you can square-cut both ends of each 4×4 rafter, but the upper end will eventually be trimmed to 16.3°, the precise angle required for creating a 9-in-12 roof pitch.

Next, make a simple rafter-layout jig for accurately marking cut lines for both the half-lap joint and the curved shape of the rafter tails. The jig (shown at left) is made from a short length of 1×4 and two ¼-in. plywood templates that are rounded on one end and mitered on the other.

4 Align the rafter-layout jig with its plywood template flush with the rafter end. Draw a line along the template to mark the angled cut line.

5 Reposition the jig so that the 1×4 is now even with the rafter end. Draw another angled line to mark the notch for the half-lap joint.

6 Use a power miter saw to cut the rafter ends to 16.3°. That way, when the roof trusses are assembled, they'll form a 9-in-12 roof pitch.

Slide the jig over the square-cut end of a 4×4 rafter. Be sure the rounded end of the plywood template is flush with the end of the rafter. Then, draw a pencil line along the template and onto the rafter tail **(1)**.

Use a jigsaw fitted with a 6-in.-long blade to cut the rounded shape into each rafter tail **(2)**. Then, clamp together the rafters and smooth all the rounded rafter tails simultaneously with a belt sander and 60-grit abrasive belt **(3)**.

Now, use the rafter-layout jig to mark the angled cut lines at the opposite end of each rafter. Start by positioning the jig so that the mitered end of the plywood template is flush with the end of the 4×4 rafter. Draw the 16.3° angled cut line along the template and onto the rafter **(4)**.

Then, slide the jig down until the end of its 1×4 is flush with the end of the 4×4 rafter. Mark another angled cut line along the template **(5)**. This second line represents the shoulder cut for the half-lap joint.

After marking cut lines onto all 10 rafters, miter-cut along the first line, trimming the rafter ends to 16.3° **(6)**. Clamp the rafters together with the mitered ends flush and evenly aligned. Adjust the circular saw to cut 1¾ in. deep and saw along the shoulder line, cutting across all 10 rafters **(7)**.

After making the shoulder cuts, prepare to cut the face of the half-lap joints. Start by snapping a chalkline across the ends of the rafters. Center the line on the 4×4s; that's 1¾ in. down from the top surface. Then, set the saw to its maximum cutting depth and saw along the chalkline and into the ends of the rafters **(8)** (see p. 46).

7 Clamp the 4×4 rafters together and adjust the circular saw to cut 1¾ in. deep. Then, saw along the shoulder line of the half-lap joints.

Snap a chalkline across the ends of the 4×4 rafters. Cut as deeply as possible along the line to establish the face of the half-lap joints.

Knock out the wood block from each half-lap joint with a hammer and chisel. Clean up the cuts by scraping the joints flat and smooth.

TOOL TIP

Sawing the bird's-mouth cuts into the roof rafters requires a bevel cut of 53°. However, some circular saws can only bevel up to 45°. Check the bevel-cutting capacity of your saw before building the roof trusses. If necessary, buy, borrow, or rent a saw that can handle a 53°-bevel cut.

The circular saw won't cut all the way through the half-lap joint, so use a hammer and chisel to knock out each block of waste wood (9). Pare away any remaining wood with the chisel.

Next, prepare to saw the bird's-mouth cut into the rafters in two passes: make the 37° seat cut first, followed by the 53° plumb cut. Begin by repositioning the rafters with their rounded tails facing up. Clamp the rafters together with the tail ends flush. Draw the lines of the bird's-mouth cuts onto the rafters. Mark the plumb cut 60 in. from the upper end of the rafter. Position the seat cut 2³⁄₈ in. down from the upper edge of the rafter at the plumb line (see the Roof Framing drawing on the facing page).

Clamp a straightedge in place and adjust the saw's bevel angle to 37°. Guide the saw along the straightedge to make the seat cuts across the rafters (10). Reposition the straightedge, set the saw's bevel angle to 53° and make the second pass to saw the plumb cuts.

Now, with each rafter cut to length and notched, you can start assembling the trusses. There are five trusses in total: three regular roof trusses that are positioned down the center of the shed, and one gable-end truss at each end of the building. Each gable-end truss receives a gable vent, so it requires a bit more framing than a regular truss.

Start by setting two rafters on a flat surface with their notched ends overlapping to form a half-lap joint. Press the joint tightly together and secure it with two 2½-in.-long structural screws.

Set the next two rafters on top of the truss you just assembled. Align the rafters with the truss below and push together the half-lap joint. Drive in two 2½-in.-long structural screws to join together the second

Saw the bird's-mouth cut in two passes: Set the saw blade to 37° and make the seat cut. Then, tilt the blade to 53° for the plumb cut.

Frame each gable-end truss for a vent. Nail the header between the vertical nailers and attach the sill. Then nail the assembled frame to the rafters.

truss. Repeat for the remaining three trusses, each time using the previous one as a template for laying out the next one.

Once you've assembled the five trusses, add the gable-vent framing to two of the trusses, thereby creating two gable-end roof trusses. Each gable-end truss is framed with four pieces of 2×4, including a 66¾-in.-long sill, two 19½-in.-long vertical nailers, and a 10¾-in. header. Cut the 2×4 framing pieces to length on the miter saw and nail them to the trusses (11).

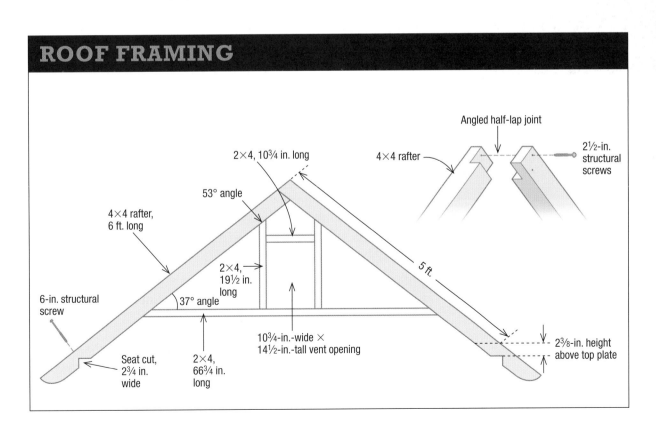

ROOF FRAMING

Angled half-lap joint

4×4 rafter

2½-in. structural screws

2×4, 10¾ in. long

53° angle

4×4 rafter, 6 ft. long

2×4, 19½ in. long

5 ft.

6-in. structural screw

37° angle

10¾-in.-wide × 14½-in.-tall vent opening

2⅜-in. height above top plate

Seat cut, 2¾ in. wide

2×4, 66¾ in. long

Fasten the gable-end truss to the top plate by driving a 6-in. structural screw into the top of the rafter and through the bird's-mouth cut.

Clamp a level to the gable-end truss, then nail a 2×3 across the roof rafters to hold the trusses perfectly plumb and properly spaced.

INSTALL THE TRUSSES

Start framing the roof by installing one of the gable-end trusses. With the help of an assistant or two, lift the truss into place and set it flush with the outside edge of the top wall plate. Attach the truss by driving a 6-in. structural screw through the top of each rafter, through the bird's-mouth cut, and into the 4×4 top plate (1). Install the remaining gable-end truss in a similar manner.

Now install the three regular roof trusses in between the two gable-end trusses. Position each truss directly over a 4×4 wall post. Secure the trusses to the top wall plate with 6-in. structural screws.

Next, mark the spacing of the trusses onto an 8-ft.-long 2×3. Lay the 2×3 across the roof trusses, a few inches down from the peak. Allow the end of the 2×3 to extend past the gable-end truss by 8 in. Nail the 2×3 to the gable-end rafter.

Clamp a level to the gable-end truss and check it for plumb (perfectly vertical). If necessary, pull or push the truss into plumb. Move down to the next truss, align it with the spacing mark on the 2×3, and then nail through the 2×3 and into the rafter (2).

Extend the previously installed 8-ft.-long 2×3 to span across the remaining roof trusses by nailing on a 40-in.-long 2×3. These 2×3s will be used later as skip sheathing for nailing on the roofing.

Secure the remaining trusses in the same manner, making sure to align each one with the appropriate spacing mark before nailing it in place. When done, nail three more rows of 2×3s across the rafters: Position one a few inches down from the peak, on the opposite side

THE RAKE SUBFASCIA

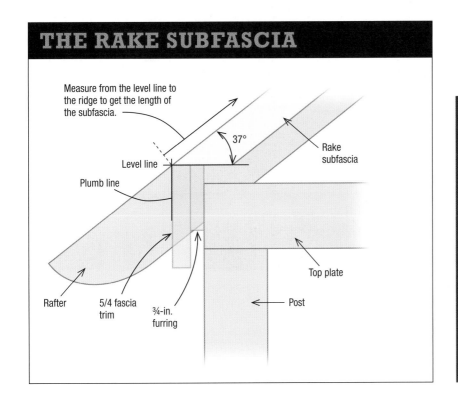

Measure from the level line to the ridge to get the length of the subfascia.

37°

Rake subfascia

Level line

Plumb line

Top plate

Rafter

5/4 fascia trim

¾-in. furring

Post

Measuring the Subfascia

To determine the length of the 2½-in.-wide rake subfascia, hold a piece of ¾-in. furring strip and 5/4 trim board against the top wall plate and alongside a roof rafter. Mark a vertical plumb line along the face of the trim board and onto the side of the rafter. Then draw a horizontal level line across the top of the trim board and furring strip.

Now measure from the ridge at the peak of the rafter down to the level line to get the length of the subfascia.

from the first 2×3. Then nail one 2×3 flush with the rafter tails along both eaves. Again, these 2×3 boards will form skip sheathing for attaching the roofing. Each row requires two 2×3s—an 8-ft. piece and a 40-in. piece—to span the entire roof frame. An alternative to using two 2×3s per row is to simply buy 12-ft.-long 2×3s and cut each one to 11 ft. 4 in.

The last step of framing the roof is to build fly rafters, which create overhangs at each gable end of the building (see the Roof Shingle Detail drawing on p. 52 and photo 2 on p. 64). Each fly rafter consists of a 2½-in.-wide rake subfascia and a 5-in.-wide rake board, which are separated by 6¾-in.-long blocks of 2×3s. However, rather than trying to cut and install all these pieces one at a time, it's much quicker and easier to partially assemble the fly rafters on the ground and then lift them into place.

Cut the two rake subfascias from 5/4 stock, mitering each end to 37°. Then cut six 6¾-in.-long blocks from a 2×3. Screw three 2×3 blocks to each of the two rake subfascias. Space the blocks as follows: Attach the first blocks 30 in. from the lower ends of each rake subfascia. Measure 17½ in. from the first blocks and attach the middle blocks.

Then, on the rake subfascia that will run along the roof plane with the clear polycarbonate panels, measure 12½ in. from the middle block and attach the third block.

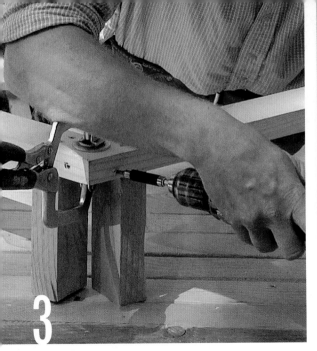

Clamp together the rake subfascias, forming a tight miter joint. Bore screw-pilot holes, then fasten the parts with 3-in. screws.

Attach furring strips to the gable-end rafters, then nail up the rake subfascia. Align the subfascia flush with the top of the rafters.

Use pocket screws to join together the two 5-in.-wide rake boards. Bore the pocket holes into the back surface so they won't be visible.

For the opposite rake subfascia, which runs along the cedar-shingle roof plane, measure up 15 in. and fasten the last 2×3 block. This precise spacing aligns the blocks with the 2×3 skip sheathing that will be installed later.

Next, clamp the two rake subfascias at the peak, forming a tight miter joint. Bore screw-pilot holes to prevent splitting, then fasten the subfascias together with two 3-in.-long exterior-grade screws (3). Now repeat the previous steps to make a second rake subfascia for the gable at the other end of the shed.

Nail 2-in.-wide furring strips to the rafters on the gable-end truss. Hold the rake subfascia in position flush with the gable-end rafter, and nail it to the furring strips (4). Use the same technique to install the second subfascia to the opposite end of the shed.

Now cut the two 5-in.-wide rake boards from 5/4 stock; miter the upper ends to 37°, but square-cut the lower ends. Clamp the rakes together at the peak and use a pocket-hole jig to bore three pocket-screw holes into the back surface. Fasten the two rake boards together by using a cordless drill to drive in three pocket screws (5).

Hold the assembled rake board in position against the subfascia with its top edge 1½ in. above the top of the subfascia. Nail through the rake board and into the ends of the 2×3 blocks protruding from the subfascia (6). Repeat these steps to fabricate and install a second rake board to the opposite gable end.

Nail the rake board to the 2×3 blocks protruding from the subfascia. Drive two 3-in.-long nails into each of the six blocks.

Mark curved cut lines onto the ends of the rake boards using the same template used earlier for scribing the ends of the rafter tails.

Next, take the jig you used earlier to mark the rounded ends of the rafter tails and hold it against the ends of the rake boards. Draw a line along the curved template and onto the rake board (7). Cut the curved shape into the rake with a jigsaw (8). Mark and cut the ends of the three remaining rake boards. Smooth the cuts with a belt sander or sanding block and 60-grit sandpaper.

Use a jigsaw to cut the rounded shape into the rake board ends. To minimize splintering, cut slowly and use a fine-tooth wood blade.

Roofing

The roof of a shed is seldom considered a particularly attractive feature, but it is in this case. With beautiful red-cedar shingles on one roof plane, and sunshine-catching clear polycarbonate panels on the other, this roof is an eye-arresting focal point.

But before you can start roofing, you must install the remaining skip sheathing. Sometimes called spaced sheathing, skip sheathing is simply rows of boards separated by a space. In this case, we used 2×3s spaced 2½ in. apart on the cedar-shingle roof plane. The polycarbonate panels require much less support, so the skip sheathing is spaced 17½ in. apart.

Skip sheathing serves three main functions: When nailed across the rafters it ties together and strengthens the roof frame. It provides solid nailing support for the roofing. And the space between the 2×3s allows air to circulate under the cedar shingles, keeping them dry and free of mold and mildew.

ROOF SHINGLE DETAIL

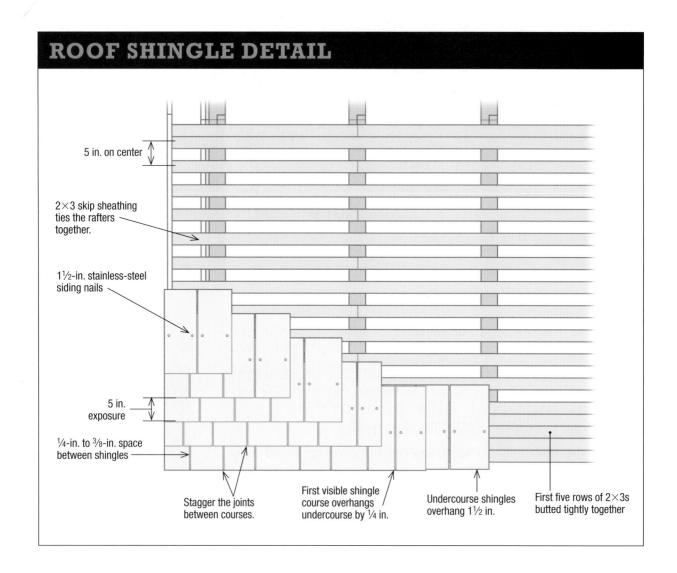

5 in. on center

2×3 skip sheathing ties the rafters together.

1½-in. stainless-steel siding nails

5 in. exposure

¼-in. to ⅜-in. space between shingles

Stagger the joints between courses.

First visible shingle course overhangs undercourse by ¼ in.

Undercourse shingles overhang 1½ in.

First five rows of 2×3s butted tightly together

SHINGLE THE ROOF

Start by nailing five 2×3s along each eave of the roof. Hold the first one flush with the end of the rafter tails and fasten it to each roof rafter with two 3½-in. nails.

Nail on the next 2×3, setting it tightly against the first one. Continue in this manner until you've installed the first five rows along both eaves. When using two 2×3s in each row to span the roof, be sure to stagger the joints between the rows. Don't align all the joints at the same rafter. The reason for butting together the first five rows of sheathing is so that when you look up under the eaves from below, you'll see a clean, finished surface.

Now, on the cedar-shingle side of the roof, nail down rows of 2×3 skip sheathing, spacing each row 2½ in. apart. Since 2×3s are 2½ in. wide, use a short 2×3 block as a gauge for quickly spacing the rows. Continue installing skip sheathing until you reach the peak (1).

HELPFUL HINT

Spacing the 2×3 skip sheathing 2½ in. apart might seem arbitrary, but it positions the 2×3 rows precisely 5 in. on center. That's the ideal distance for supporting the red-cedar shingles, which are installed with a 5-in. exposure to the weather.

To support the red-cedar shingles, nail 2×3 skip sheathing to the rafters. Space the rows 2½ in. apart and fasten them with 3½-in. nails.

On the opposite side of the roof, nail down three rows of 2×3 skip sheathing to support the lightweight clear polycarbonate panels.

Once you've completed the skip sheathing on the cedar-shingle roof plane, move to the other side of the roof. Measure up 17½ in. from the 2×3s nailed along the eave and install the first 2×3. From the first 2×3, measure up another 17½ in. and nail on the second row. Position the third row 12½ in. up the roof from the second row (2). (See the Clear Roof Panel Detail drawing on p. 55.)

With the skip sheathing installed, you can begin shingling the roof. Start by screwing a temporary 2×4 spacer to the lowermost 2×3 fastened across the rafter tails at the eave. Hold the top edge of the 2×4 spacer flush with the top surface of the 2×3 in the first row. Then, screw a 1×4 ledger board to the 2×4 spacer, allowing its upper edge to extend ¾ in. above the spacer.

The 2×4 spacer creates the necessary 1½-in. overhang for the very first course of cedar shingles, known as the undercourse. And the raised lip of the ledger serves two purposes: It allows you to lay out the shingles without having them slide off the roof, and it holds the shingles in a straight line for nailing.

Lay several undercourse shingles onto the roof, allowing them to rest against the ledger. Adjust the first shingle so that its edge overhangs the rake trim by ¾ in. Using 1½-in.-long nails, fasten the shingle to the 2×3 sheathing. Position the nails 6½ in. up from the butt (thick) edge of the shingle, so that the shingle course above will cover the nail heads by 1½ in. Set the next shingle into place, ¼ in. to ⅜ in. from the first shingle, and nail it. Continue nailing down shingles to complete the undercourse (3).

Attach a 2×4 spacer and 1×4 ledger to the eave, then nail on the very first row of shingles, which is known as the undercourse.

TOOL TIP

Hand-nailing cedar shingles is a peaceful, pleasant job, but using a pneumatic coil roofing nailer is quicker and easier. Don't own a roofing nailer? No problem. Pneumatic tools and air compressors are available for rent at most tool-rental shops.

Nail the first visible course of shingles directly over the undercourse shingles. Position the nails 6½ in. from the butt edge.

Make a jig to hold the shingles in place for nailing. Leave a gap between each shingle and stagger the joints from one course to the next.

HELPFUL HINT

Here's an easy way to gauge the ¼-in. overhang of the first course of shingles: Hold a carpenter's pencil—which is ¼ in. thick—against the undercourse shingle. Slide down the first-course shingle until it's flush with the pencil, then nail it in place.

Once you've installed the undercourse, unscrew and remove the ledger board and 2×4 spacer. Now install the first visible course of shingles right on top of the undercourse shingles **(4)**. Allow the first-course shingles to extend past the butt edge of the undercourse by ¼ in.

Space the shingles ¼ in. to ⅜ in. apart and stagger the seams between the first course and the undercourse. Install the subsequent shingle courses in a similar manner, making certain that each course is exposed 5 in. to the weather.

To speed up the installation, make a shingling jig, which consists of two short vertical 1×4 hangers screwed to a long horizontal 1×4 ledger. Lay the jig on the roof with the top edge of the ledger 5 in. above the butt edge of the previously installed shingle course. Then, tack-nail the vertical hangers to the roof. Now, as you lay out the shingles across the ledger, they'll automatically be positioned with the desired 5-in. exposure **(5)**.

INSTALL THE POLYCARBONATE PANELS

Take the 2×4 spacer and 1×4 ledger board used earlier, and screw them to the eave where the clear polycarbonate roof panels will be installed. The spacer and ledger will perform the same functions as before: create the necessary 1½-in. overhang and provide a raised lip for holding the roofing material in place.

Next, stack and clamp together the polycarbonate panels. Cut them all to length at the same time using a circular saw fitted with a thin-kerf, carbide-tipped wood blade. To produce the smoothest, cleanest cut, the roof-panel manufacturer recommends mounting the blade backward on the saw **(1)**.

1 **To save time,** gang-cut all the poly-carbonate roof panels at once. For the cleanest cut, mount the blade backward on the circular saw.

2 **Mark the locations of the skip sheathing** onto the roof panels. Then, bore ¼-in.-dia. pilot holes through the panels at each mark.

3 **Fasten the polycarbonate roof panels** with rubber-gasketed screws. Don't over-tighten the screws or you'll crush the corrugations.

With the polycarbonate panels still clamped together, mark the positions of the 2×3 skip sheathing. Then, at each mark bore a ¼-in.-dia. screw-pilot hole through the raised portion of the corrugated panels (**2**). (Drilling holes in the troughs will invite leaks.)

Set the first polycarbonate panel into place with its bottom end resting on the ledger board. Slide the panel over so that it extends ¾ in. past the rake. Fasten the panel to the 2×3 skip sheathing with 2-in.-long rubber-gasketed screws (**3**). Be careful not to overtighten the screws or you'll deform the corrugated panel and compress the rubber gasket so much that it'll prevent the panel from expanding and contracting.

SAFETY FIRST

When using a circular saw to cut the polycarbonate roof panels, be sure to wear hearing protection. The panels cut easily, but make an enormous, ear-splitting racket.

CLEAR ROOF PANEL DETAIL

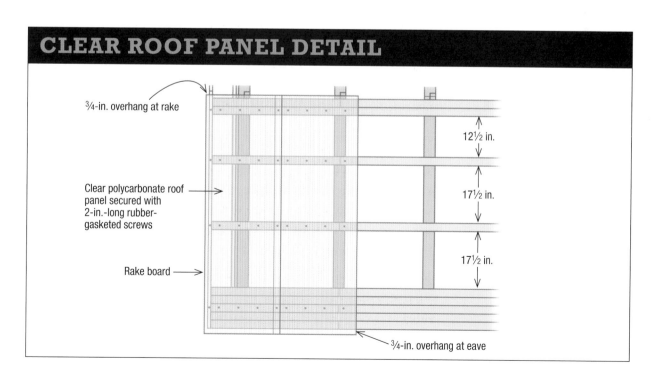

¾-in. overhang at rake

Clear polycarbonate roof panel secured with 2-in.-long rubber-gasketed screws

Rake board

12½ in.

17½ in.

17½ in.

¾-in. overhang at eave

4

To help seal out rain, overlap the panels by at least one full trough. Rip the last panel to fit, making sure it extends ¾ in. past the rake.

5

Make a continuous ridge cap by screwing a 4¼-in.-wide cedar board to a 1×4 cedar board. Join the boards with 1⅝-in. decking screws.

6

Set the ridge cap over the roof peak. Bore pilot holes every 18 in. or so along both sides of the cap, then fasten it with 2½-in. screws.

As you work your way across the roof, overlap the corrugated panels by one full trough **(4)**. You'll have to rip the last panel to fit. Again, use a circular saw with a backward-facing blade.

After installing all the polycarbonate panels and cedar shingles, make a ridge cap to fit along the peak—or ridge—of the roof. Rip a cedar 1×6 down to 4¼ in. wide. Then fasten it to a cedar 1×4 with 1⅝-in. galvanized screws, forming the L-shaped cap **(5)**. Set the cap onto the ridge, bore pilot holes, then secure it with 2½-in. galvanized screws **(6)**.

Windows and Door

The shed has a single 30-in.-wide wood entrance door and seven windows, each measuring 21½ in. wide by 57 in. tall. That's a lot of windows—and a lot of glass—for a building this size, but we wanted to let in as much natural light as possible.

However, with so many windows there isn't much interior wall space left for mounting shelves, hanging storage cabinets, or installing a potting bench or work surface. To gain the most natural light without sacrificing all the wall space, we placed just one window on the rear gable-end wall and installed no windows at all along the left-hand sidewall.

INSTALL THE WINDOWS

Before fitting the windows into the walls, you must reduce the width of the rough openings. Start by ripping 1-in.-thick by 1½-in.-wide wood strips from 2-by stock. Then, crosscut the strips to length and nail them to the sides of each rough opening **(1)**. The 1-in.-thick strips will provide solid support for fastening the windows.

Nail 1-in.-thick vertical wood strips to each window rough opening. The strips provide support for the windows' mounting flanges.

Use a level to plumb up each window. Make the spaces between windows even and parallel. Otherwise, you'll have to cut tapered pieces of trim to fit.

Plumb the door, then secure it to the rough opening by screwing through the jambs and into the 4×4 wall posts on either side.

Set the first window onto the horizontal 2×4 girt, which acts as a rough sill. Tilt the window into the rough opening and check it for plumb and level. If necessary, have a helper slip shims between the window and wall framing.

Once satisfied with the window's position, drive 1½-in. pan-head screws through the slots in the nailing flanges and into the 1-in.-thick wood strips. Install the remaining windows in a similar manner, always making sure to use a long level to align each window perfectly plumb and level (2).

HANG THE DOOR

For this shed, we bought a fir door and a separate doorjamb kit. We assembled the jambs, mounted the hinges, and fit the door into the jamb. An easier—though more expensive—option is to install a prehung door.

Regardless of whether you use a jamb kit or prehung door, start by sealing the bottom of the door and side jambs with primer and paint. If you plan on staining the door, seal the bottom surfaces with exterior-grade polyurethane varnish. Sealing the end grain helps block out moisture, which can lead to rot.

Set the door into the rough opening with its jamb protruding ¾ in. from the 4×4 wall posts. This positioning will align the doorjamb flush with furring strips that you'll install later. Slip shims between the side jambs and wall posts. Check the door for plumb and be sure there's an even gap between the door edges and the jambs. Secure the door by driving 2½-in. decking screws through the side jambs and into the wall posts (3). Trim any protruding shims flush with the wall framing.

Fasten 1×4 furring strips to each wall post and corner post. Note that two furring strips are nailed to the 6-in. face of the 4×6 corner posts.

Nail a continuous sill to the tops of the furring strips. The 3-in.-wide sill separates the top half of the walls from the bottom half.

Siding and Trim

With an 80-in.-tall door and seven 57-in.-tall windows there's not a whole lot of siding to install on this shed. But that doesn't mean what little siding there is can't have a significant impact on the architectural style and beauty of the building. Here, we installed cedar bevel siding (a.k.a. clapboards) on the lower half of the walls and board-and-batten siding on the upper half.

The two siding types are separated by a continuous sill and accented by wide white-painted trim boards installed along the bottom and top of each wall and at each shed corner. In fact, before nailing on any siding, you must first install the trim.

INSTALL THE LOWER TRIM

To keep the bevel siding in the same plane as the board-and-batten siding, it's necessary to install ³/₄-in.-thick furring strips to the lower half of the shed. Cut 1×4 furring strips to length and nail one to each wall post and corner post. Position the top end of each furring strip precisely 2⅛ in. down from the top edge of the horizontal 2×4 girts (**1**).

Next, make the continuous sill that runs around the building, separating the top half of the wall from the bottom half. You'll need to cut five sills in total: one long sill for each sidewall, a slightly shorter sill for the rear gable-end wall, and then two very short sills for the left and right sides of the entrance door.

Join the sill pieces at each corner with a miter joint. Bore a pilot hole first, then screw each joint together with a single 3-in. screw.

Use the tablesaw to bevel-rip 45° miters into the corner boards. Then, fasten the boards together with 1⅝-in.-long trim-head screws.

Start by using a tablesaw to rip 2×4s down to 3 in. wide. Then, tilt the blade to 10° and cut a bevel into the top of each sill. The beveled surfaces help drain away rainwater, but they also look better than flat surfaces.

Cut the sills to length, mitering the ends at 45° to form a miter joint at each corner post. Set the sills on top of the furring strips, and nail at a slight angle through the sill and furring strips and into the wall posts and corner posts (2). Fortify each corner joint by driving a single 3-in.-long screw through the edge of the miter joint (3). To keep the screws from splitting the sills, be sure to bore pilot holes first.

Next, fabricate four corner boards for the lower half of the shed. Make each corner board out of two pieces of 5/4 primed trim. Begin by cutting the trim boards to length. Then, tilt the tablesaw blade to 45°, and for each of the four corner boards, bevel-rip one trim board to 6¾ in. wide and another to 8¾ in. wide. After cutting all eight trim pieces, you can assemble the four corner boards.

Glue and clamp one 6¾-in.-wide trim board to an 8¾-in.-wide trim board. Adjust the mitered corner joint for a tight fit, and fasten the boards together with 1⅝-in. exterior-grade trim-head screws (4). Repeat to assemble the remaining three corner boards.

Set each corner board over a corner post and tight up against the underside of the continuous sill. Attach the corner board with 2½-in. ring-shank nails (5).

Hold the assembled corner board on the corner post and tight to the bottom of the sill. Secure the corner board with ring-shank nails.

Cut a 6-in.-wide skirt board to fit snugly between the corner boards. Fasten the skirt to the furring with 2½-in.-long ring-shank nails.

Nail a wood drip-cap to the top edge of the skirt boards. The drip-cap has a beveled surface, which will help drain away rainwater.

Next, cut a 5/4 by 6-in. skirt board to fit in between each corner board. The skirt boards cover up the floor frame and run horizontally along the base of the shed walls. Nail the skirt boards in place with 2½-in. ring-shank nails (6). Then, cut wooden drip-cap molding to length and fasten it to the top of each skirt board with 1½-in. nails (7).

NAIL ON THE BEVEL SIDING

Measure the distance between corner boards and cut a length of cedar bevel siding about ⅛ in. longer. Bow out the middle of the siding, then set it between the corner boards and on top of the drip-cap. Push in on the siding to snap it tightly into place (8). Nail the siding to the wall posts and corner posts with 2-in. nails.

Continue installing cedar bevel siding up the wall, maintaining 4-in. exposure to the weather (9). When using two lengths of siding to span the wall, be sure the butt joint between the pieces is centered on a 4×4 wall post. Rip the last row of siding to fit beneath the continuous sill, making sure to maintain the 4-in. exposure. Now cut and install bevel siding on the lower half of the remaining three shed walls.

Cut the cedar bevel siding slightly longer than the distance between the corner boards. Snap it into place and fasten with 2-in. nails.

When installing the cedar siding, measure frequently to maintain a 4-in. exposure to the weather. Center all butt joints over a post.

INSTALL THE BOARD-AND-BATTEN SIDING

The traditional board-and-batten siding installed on the upper half of the shed is composed of 1×10 boards and 1×3 battens. The boards are spaced 1½ in. apart along both gable-end walls and ¾ in. apart along the sidewalls. It's important to leave these spaces so the boards have room to expand and contract. The battens conceal the spaces between the boards and are nailed to the wall framing, not to the boards. Again, that's so the boards are free to move without splitting.

Begin by installing a 1×10 board on the center of the rear gable-end wall. It's critical that this very first board be perfectly plumb. If it's not, it'll throw off all the subsequent boards. Hold the 1×10 in place on the wall's centerline, adjust it to plumb with a 4-ft. level, and then fasten the board with 2½-in. ring-shank nails (1) (see p. 62).

Rip a 1½-in.-wide strip of wood for use as a spacer. Hold the spacer against the edge of the first 1×10, set the next board against the spacer, and then nail the second board. Pull the spacer from between the boards and repeat, using it to maintain consistent 1½-in.-wide spacing between the 1×10 boards (2) (see p. 62).

After installing the last 1×10 board on the rear gable-end wall, nail a horizontal trim board across the upper wall section and level with the top of the window. Then, cut and install 1×10 boards to the left-hand sidewall, which is the wall without windows. Remember to space the boards only ¾ in. apart.

HELPFUL HINT

Whenever you cut a length of primed siding or trim, it's important to seal the fresh cut with a coat of primer to block out moisture. You can brush on the primer, but spray primer goes on faster and dries quicker.

Start the board-and-batten siding centered on the rear gable wall. Plumb and nail the first 1×10 board, then work out in both directions.

As you nail up the 1×10 boards, maintain the 1½-in. expansion space by holding a 1½-in.-wide spacer strip between the boards.

Now cut 1×3 battens to length and nail them over the spaces between the 1×10 boards **(3)**. Check each batten for plumb and nail it to the wall framing.

Next, cut a fascia for the left-hand sidewall from a length of 5/4 by 6-in. trim board. Hold the fascia close to the top of the sidewall and mark the positions of the rafter tails onto the fascia. Use a jigsaw to cut the notches into the fascia on the marked lines **(4)**. Slide the fascia up into place, fitting it around all the rafter tails, then secure it with 2½-in. nails. Note that you don't have to notch the fascia for the opposite sidewall. Simply cut individual pieces of 5/4 trim to fit between the rafters.

Continue installing the remaining 5/4 trim boards, including those between and above the windows and around the doorframe. Then build or buy louvered vents to fit into the framed openings near the top of each gable-end wall. Nail the vents to the siding **(5)**.

Cover the spaces between the 1×10 boards with vertical 1×3 battens. Center and plumb the battens and nail them to the wall framing.

Use a jigsaw to notch the fascia board to fit around the rafter tails at the top of the left sidewall; that's the wall without windows.

Fasten a louvered vent to the framed opening at the top of each gable-end wall. The vents are needed to let hot air escape the shed.

1

2

3

1. A bank of four 57-in.-tall double-hung windows is installed in the right sidewall directly below the cedar-shingle roof.

2. Fly rafters form gable overhangs at each end of the roof. Note how the clear panels extend onto the overhang, creating patterns of light and shadows.

3. The rounded tails cut into the ends of the 4×4 rafters are exposed along the sidewalls, directly below the overhanging eave of the roof.

4. The shed's timber frame is exposed on the interior. Note that the 4×4 rafters and 2×3 skip sheathing support the clear roof panels.

5. The left sidewall has no windows, creating space for shelves, but plenty of sunlight floods into the shed through the clear roof panels.

6. The rear end of the shed has just one window to provide some additional sunlight but still leave enough wall space for installing shelves.

3

BOARD-AND-BATTEN SHED

a cellular PVC product that's impervious to cracking, rotting, and water damage. Cellular PVC costs more than wood, but in the long run you'll save time and money by never having to paint, sand, scrape, repair, or replace the sash or trim. (To order building plans for the Board-and-Batten Shed, see Resources on p. 214.)

Foundation and Floor Frame

Since the footprint of this shed is only 100 sq. ft., the building inspector allowed us to use an on-grade foundation. That saved us the trouble of digging down to the frost line. However, building codes vary widely across the country, so be sure to check with your local building department before starting construction.

This particular foundation consists of nine solid-concrete blocks laid out in three parallel rows with three blocks per row. Each block measures 4 in. thick by 8 in. wide by 16 in. long. The floor frame is built on top of the blocks and is then covered with ¾-in. plywood. To prevent rot and insect damage, be sure to build the floor frame with pressure-treated lumber.

SET THE BLOCKS AND FRAME THE FLOOR

Arrange the nine solid-concrete foundation blocks in three parallel rows. Make each row 9 ft. 11 in. long, and space the first and third row 9 ft. 11 in. apart, as measured from outer edge to outer edge. Center the middle row over blocks between the two outer rows. Lay a long, straight 2×4 on edge across one row of blocks. Then, set a 4-ft. level on top of the 2×4 and check the blocks for level (1). If the ground is sloping, shim up the low blocks with pieces of solid, weather-resistant material, such as additional 2-in.- or 4-in.-thick solid-concrete blocks, pressure-treated wood, composite lumber, or asphalt roof shingles.

Use a long, straight 2×4 and 4-ft. level to make sure that each row of three solid-concrete foundation blocks is perfectly level.

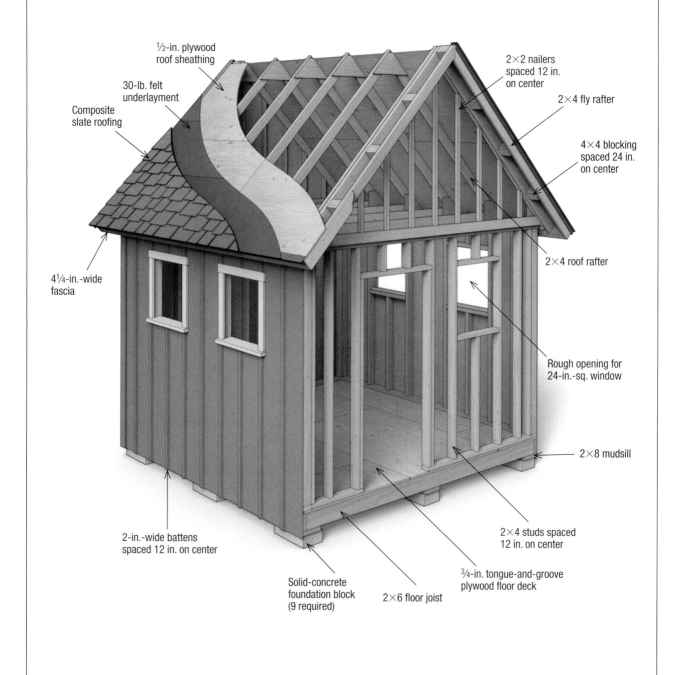

½-in. plywood roof sheathing

30-lb. felt underlayment

Composite slate roofing

4¼-in.-wide fascia

2×2 nailers spaced 12 in. on center

2×4 fly rafter

4×4 blocking spaced 24 in. on center

2×4 roof rafter

Rough opening for 24-in.-sq. window

2×8 mudsill

2-in.-wide battens spaced 12 in. on center

Solid-concrete foundation block (9 required)

2×6 floor joist

¾-in. tongue-and-groove plywood floor deck

2×4 studs spaced 12 in. on center

2 Cut 2×6 floor joists to fit in between the rim joists. Set the floor joists down on top of the mudsills, spaced 16 in. on center.

3 Push down on the floor joist to hold it flush with the rim joist. Drive two 3-in.-long screws through the rim joist and into the floor joist.

4 Fasten each floor joist to the center mudsill with a single 3-in. screw. Check to be sure the joists are 16 in. on center before fastening.

Repeat to level the other two rows of blocks. Then level across the rows, making sure all three rows of foundation blocks are level with each other.

Cut three 10-ft.-long mudsills from pressure-treated 2×8s. Then cut two 10-ft.-long rim joists from pressure-treated 2×6s. Fasten each 2×6 rim joist to a 2×8 mudsill with 3-in. galvanized decking screws, forming an L-shaped assembly. Be sure to drive the screws through the 2×8 and into the edge of the 2×6 rim joist.

Set the L-shaped assemblies on top of the outer two rows of foundation blocks. Position each assembly with the 2×8 mudsill down flat against the concrete blocks and the 2×6 sticking straight up. Lay the remaining 2×8 mudsill across the row of blocks in the center of the foundation. (This center 2×8 sill doesn't require a 2×6 rim joist.)

Next, cut nine floor joists from pressure-treated 2×6s. Make each joist 9 ft. 9 in. long. Set the floor joists on edge in between the rim joists and on top of the mudsills (2). Space the floor joists 16 in. on center. Secure each end of every joist by driving two 3-in. decking screws through the rim joist and into the ends of the floor joists (3). Tack the center of each floor joist to the middle mudsill with a single 3-in. screw driven down at an angle (4).

INSTALL THE FLOOR DECK

Cover the floor frame with ¾-in. tongue-and-groove plywood (1) (see p. 72). Keep the plywood flush with the outer edges of the floor frame, and align all end-butt joints over the center of a floor joist. Secure the plywood to the joists below with 2-in. galvanized decking screws. Space the screws 8 in. to 10 in. apart and be sure to drive them slightly below the surface of the plywood. If any screw heads are left sticking up even just a little, they'll create a tripping hazard.

TOOL TIP

A cordless impact driver provides a fast, nearly effortless way to drive screws, but be sure to have two battery packs on hand. Then, when one battery runs out of power, you can pop it into the charger, grab the second battery, and keep working. Most cordless-tool batteries can be recharged in one hour or less.

Anchoring the Shed

In some municipalities, sheds built with on-grade foundations must be secured to the ground to prevent them from being blown over. The likelihood of that happening to a wooden shed is extremely low, but this building code exists because metal sheds blow over quite easily and can cause extensive damage and injuries. The code applies to all sheds, regardless of size or type. So, if the local building department requires you to tie down your shed, use ground anchors.

A ground anchor is simply a sharpened metal spike attached to a length of steel cable. To install the anchor, screw the free end of the cable to a mudsill or floor joist, as shown in the photo above. Then use a long steel rod and sledgehammer to drive the spike deep into the ground. (Note that some anchors have spiral shafts that you twist into the ground.)

You can typically satisfy the building code with two ground anchors—installed at opposite corners—but check with the building inspector to be sure.

1 **Lay ¾-in. tongue-and-groove plywood** over the floor frame. Tap the sheets together and be sure that all end joints are centered on a joist.

FLOOR FRAME

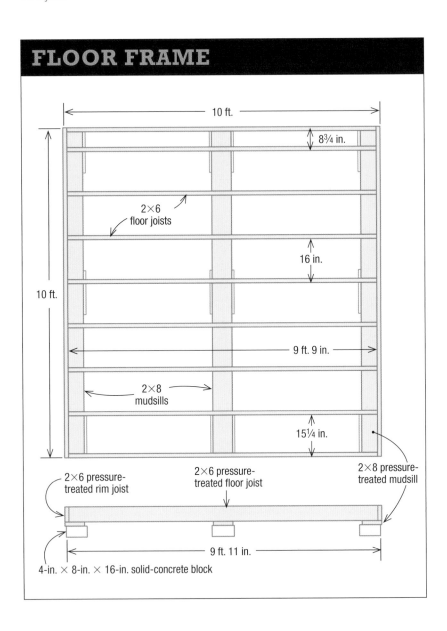

10 ft.

8¾ in.

2×6 floor joists

16 in.

10 ft.

9 ft. 9 in.

2×8 mudsills

15¼ in.

2×6 pressure-treated rim joist

2×6 pressure-treated floor joist

2×8 pressure-treated mudsill

9 ft. 11 in.

4-in. × 8-in. × 16-in. solid-concrete block

1 Temporarily screw short 2×4 blocks down to the floor deck to act as stop blocks for positioning and assembling the roof trusses.

2 Cut the ½-in. plywood gusset plates to size on the miter saw. Set the saw's miter angle to 45° to match the shed's 12-in-12 roof slope.

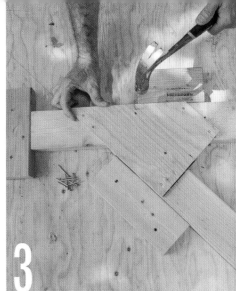

3 Set the gusset plate flush with the outer edge of the rafter and bottom edge of the chord; secure it with 1½-in.-long roofing nails.

Build Roof Trusses

The roof of this shed has a 12-in-12 slope, meaning that each roof plane is at a 45° angle. This roof slope was chosen because it looks appropriate for this particular size building, but also because it makes framing the roof much easier: All angle cuts are made at 45°. However, rather than stick-building the roof frame one board at a time, we simplified the process by preassembling roof trusses, using the flat floor deck as a giant workbench.

The roof frame is made up of six roof trusses, including two gable-end trusses and four standard trusses. Each truss consists of two 2×4 angled roof rafters and a horizontal 2×4 chord. The parts are held together with ½-in. plywood gusset plates nailed across the joints. The gable-end trusses also have 2×2 studs for nailing on the plywood siding, and one gable-end truss has a rough opening to receive a diamond-shaped window.

ASSEMBLE THE TRUSSES

Begin by cutting twelve 2×4 roof rafters to 7 ft. 10⅜ in. long. That's two rafters for each roof truss. Trim the top end of each rafter to 45° and square-cut the opposite end. Then cut six 2×4 chords to 10 ft. long. Trim the upper corners of each joist to 45°, as shown in the Roof Truss drawing on p. 74.

Next, lay a truss onto the plywood floor deck using two rafters and one chord. Fit the two rafters tightly together at the peak to form a 90° angle. Lay the chord between the rafters, equidistant from the square end of each rafter.

4 Cut a triangular gusset plate to fit at the peak of the truss. Hold it flush with the outer edges of the two rafters, then nail the gusset in place.

HELPFUL HINT

You can use standard square-edge plywood for the floor deck, but tongue-and-groove plywood locks together, creating a much stronger, more rigid floor. This is especially important if you plan to use the shed to store woodworking machines, a lawn tractor, or other heavy items.

5 Frame the gable-end trusses with 2×2 studs. Cut a long center stud to fit from behind the plywood gusset plate to the 2×4 chord.

6 Use a layout square to square up the center stud with the chord. Then, fasten the lower end of the stud with a 3-in. decking screw.

7 Cut 2×2 studs to fit along each gable-end truss. Space the studs 12 in. apart and fasten each end with a single 3-in. decking screw.

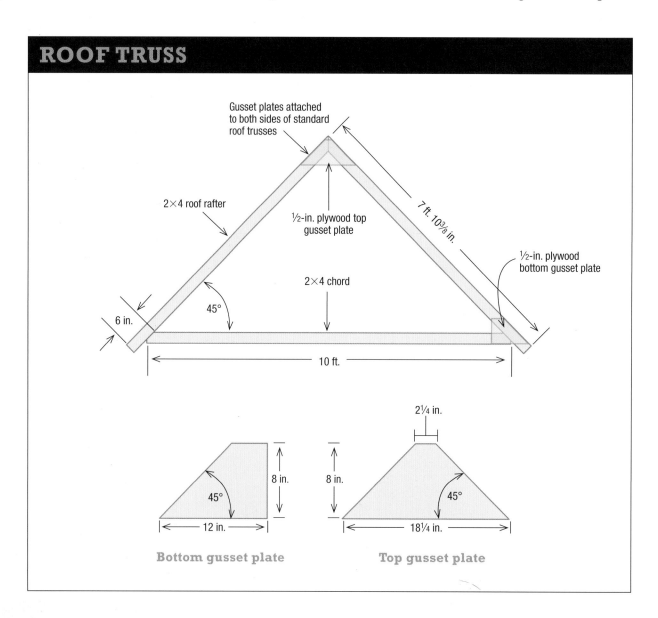

ROOF TRUSS

Gusset plates attached to both sides of standard roof trusses

2×4 roof rafter

½-in. plywood top gusset plate

7 ft. 10³⁄₈ in.

½-in. plywood bottom gusset plate

2×4 chord

45°

6 in.

10 ft.

Bottom gusset plate

8 in.

45°

12 in.

Top gusset plate

2¼ in.

8 in.

45°

18¼ in.

8

Screw together 2×4s to form a rough opening for the 20-in.-sq. diamond-shaped window that's installed in the rear gable-end wall.

9

Fasten a 2×4 mounting plate to the chord on each gable-end truss. Hold the plate flush with the chord and attach it with 3-in. screws.

Roof Truss Trio

Here are photos of the three different types of roof trusses required to frame the shed roof.

Each of the four standard roof trusses consists of just three 2×4 members: two rafters and a bottom chord.

The front gable-end truss has 2×2 studs spaced 12 in. on center, which provide solid nailing for the siding and battens.

The rear gable-end truss is framed with studs, but also has a rough opening for the 20-in.-sq. diamond-shaped window. If you decide not to install a window, then the two gable-end trusses would be identical.

And remember that plywood gusset plates are installed to both sides of the four standard roof trusses, but to only the inside of the two gable-end trusses.

Now cut ten 2×4 stop blocks, making each one 8 in. to 10 in. long. Set the stop blocks into place against the ends and edges of the two rafters and along the edges of the chord. Fasten the stop blocks to the floor deck with 3-in. screws **(1)** (see p. 73).

Use a miter saw, set to 45°, to cut the 1/2-in. plywood gusset plates **(2)** (see p. 73). Gusset plates hold together the truss parts and are installed in three places: at each end of the chord and at the peak where the two rafters come together. Note that gusset plates are nailed to both sides of the four standard roof trusses, but to only the inside of the two gable-end trusses.

Secure the gusset plates to the truss with 1 1/2-in.-long roofing nails **(3)** (see p. 73). Install the triangular-shaped gusset plates across the miter joint at the peak of the each roof truss **(4)** (see p. 73). Once all the gussets are nailed in place, remove the truss from the floor deck and prepare to assemble the next truss. Set two more rafters and another chord against the stop blocks and nail on the gusset plates. Repeat until you've assembled six trusses.

Now add 2×2 studs to each of the two gable-end trusses. Slip the long center stud into place behind the gusset plate at the truss peak and secure the upper end of the stud with 1 1/2-in.-long roofing nails **(5)**. Fasten its lower end with a single 3-in.-long screw **(6)**. Continue to install 2×2 studs across the truss, spacing them 12 in. on center **(7)**.

When framing the gable-end truss at the rear of the shed, cut 2×4s to create a 16-in.-sq. rough opening for the diamond-shaped window. Fasten the 2×4s together with 2-in. screws **(8)**.

Once you've framed both gable-end trusses, attach a 2×4 mounting plate to the chord. Hold the plate flush with the bottom edge of the chord and secure it with 3-in. screws spaced 10 in. apart **(9)**.

Screw ⅜-in.-thick rough-sawn plywood siding to the gable-end trusses. Space the 1⅝-in. decking screws about 10 in. apart.

Locate the window's rough opening and drill a ¾-in.-dia. access hole through the plywood siding to the inside of the opening.

Rout the plywood from the rough opening with a ball-bearing piloted flush-cutting bit. Move the router clockwise around the opening.

Screw down the plywood around the perimeter of the rough opening, then carefully set the diamond-shaped window into place.

TOOL TIP

You can cut all the roof-truss parts with a portable circular saw, but a power miter saw is quicker, safer, and more accurate. If you don't own a power miter saw, buy, borrow, or rent one. It'll save you a lot of time and trouble in the end.

ATTACH THE PLYWOOD SIDING

After assembling the six roof trusses, stack the four standard trusses off to one side. Unscrew and remove the 2×4 stop blocks from the floor deck. Place the rear gable-end truss on the floor deck with its outer surface facing up. (The plywood gussets should be on the underside, facing the floor deck.)

Now cut ⅜-in.-thick rough-sawn plywood to match the 45° angle of the truss. Secure the plywood to the truss with 1⅝-in. decking screws, spaced 8 in. to 10 in. apart **(1)**. After covering the entire truss in rough-sawn plywood siding, drill a ¾-in.-dia. access hole through the siding and into the window's rough opening **(2)**.

Next, use a router fitted with a ball-bearing piloted flush-cutting bit to cut the plywood from the rough opening. Adjust the router's depth of cut to about ¾ in. Set the bit into the access hole, turn on the router, and move the router in a clockwise direction around the rough opening. As you advance the router, keep the bit's ball-bearing pilot pressed against the rough opening to trim away the plywood **(3)**.

After routing the window opening, screw down the plywood around the perimeter of the rough opening and set the window sash into place **(4)**. Secure the sash to the truss with 2½-in.-long trim-head screws.

Now attach rough-sawn plywood siding to the front gable-end truss, a process that goes much faster because it has no window opening.

1

The first step to building the gable overhang is to screw a 2×4 rafter to short 4×4 blocks. Fasten each block with three 3-in. screws.

BUILD THE GABLE OVERHANGS

The roof of this shed extends several inches beyond each gable end, creating an architectural element known as a gable overhang. Now this construction detail isn't absolutely necessary; in fact, most sheds don't have gable overhangs. However, gable overhangs, which are also called fly rafters, create interesting shadow lines and offer a modicum of protection to the gable ends. But the best reason for adding gable overhangs is much simpler: a shed just looks better with them.

For each gable-end truss, cut seven 5-in.-long 4×4 blocks and four 2×4 rafters. Make the four gable-overhang rafters identical to the roof rafters cut earlier for the trusses. Fasten three blocks each to two gable-overhang rafters. Space the blocks 24 in. on center and attach each one with three 3-in.-long decking screws (1). Screw the remaining 4×4 block to the mitered top end of one rafter, allowing half of the block to extend past the rafter. The block's exposed half will be covered by the opposite rafter.

Once all the blocks are attached, turn over the assemblies so that both gable-overhang rafters are flat against the truss and the 4×4 blocks are sticking straight up. Align the edges and ends of the gable-overhang

Flush-Cut Routing

To rout window openings in the plywood siding, you'll need a router with a 2-hp or larger motor and a flush-cutting bit that's equipped with a ball-bearing pilot. The bearing rolls along the interior of the rough opening, trimming the plywood perfectly flush with the opening.

Using a router to cut window openings is quick, easy, and very accurate. In fact, this method was used to cut all the window openings in this shed. However, if you don't own a router, you can cut the openings with a jigsaw, reciprocating saw, or circular saw. Just be careful not to cut past the opening and through the framing.

SAFETY FIRST

A router is one of the most useful and versatile power tools ever made. It's also one of the loudest. Always wear hearing protection when routing to help silence the high-pitch scream of the router motor.

Flip over the gable-overhang rafter and use a layout square to make sure that its end is aligned perfectly flush with the rafter on the truss.

Nail 2-in.-wide battens strips to the plywood siding. Space the battens 12 in. on center and fasten them with 2½-in.-long galvanized nails.

rafters with the roof rafters attached to the truss (2). Fasten the gable-overhang rafters to the truss with 3-in. decking screws spaced about 12 in. apart.

Next, use a tablesaw to cut 2-in.-wide batten strips from pine 1×8s or 1×10s. After ripping the battens, tilt the sawblade to 10° and run the battens through the saw again to chamfer the edges. Chamfering isn't necessary, but it does give the battens a more interesting, finished appearance. Fasten the batten strips to the gable-end truss with 2½-in.-long galvanized nails (3). Space the battens 12 in. on center, so that the nails go through the plywood siding and into the 2×2 studs.

After nailing on all the batten strips, complete the gable overhang by placing the remaining 2×4 rafters on top of the 4×4 blocks (4). Use a framing square to ensure the ends of the rafters are flush. Then, fasten the top rafter to the blocks with 3-in.-long decking screws (5).

Next, cut two 4¼-in.-wide rake boards from pine 1×6s. Square-cut both ends of each rake and make them about 10 in. longer than the gable-overhang rafters. Set the first rake board into place, keeping it flush with the upper edge of the 2×4 gable-overhang rafter (6). Nail the rake to the rafter with 2-in. galvanized nails (7). Nail the other rake board to the opposite rafter.

Repeat the previous steps to build a gable overhang onto the remaining gable-end truss.

Place a 2×4 rafter on top of the 4×4 blocks. Check to be sure its mitered end is centered on the 4×4 fastened at the peak of the truss.

Fasten the upper rafter to the row of 4×4 blocks. Drive three 3-in.-long decking screws through the rafter and into each 4×4 block.

Wall Framing

The walls of this 10-ft. by 10-ft. shed are conventionally framed out of 2×4s, meaning that all the parts—top and bottom plates, wall studs, sills, headers, and trimmers—are cut from standard construction-grade 2×4s.

The one unconventional aspect is that the wall studs are spaced 12 in. on center, not 16 in., which is much more common. Here, the studs weren't spaced closer together to add strength to the structure; studs spaced 16 in. on center would've been plenty strong enough. The studs were spaced 12 in. on center simply to provide nailing support for the batten strips.

Each wall frame is screwed together on the floor deck and then the plywood siding, batten strips, and window trim are installed prior to tipping up the walls. Prefabricating each wall on the floor deck is easier and quicker than waiting until the wall frame is erected.

Set the 4¼-in.-wide rake board into place on top of the 2×4 gable-overhang rafter. Hold the top edge of the rake flush with the 2×4 rafter.

Use 2-in. nails to attach the rake board to the rafter. Note that the rake is cut long to extend several inches past the end of the 2×4 rafter.

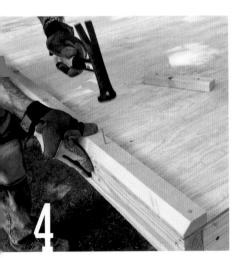

1 Space the 2×4 studs 12 in. on center. Fasten each stud with two 3-in.-long decking screws driven through the top and bottom wall plates.	**2** To determine if the wall frame is square, measure the opposite diagonals. If the two dimensions are identical, the frame is square.

3 Cover the exterior of the wall frame with ³⁄₈-in.-thick rough-sawn plywood. Cut the plywood 8½ in. longer than the wall height.

4 Nail a 2-in.-wide pine starter strip to the top of the sidewall. The chamfered edge of the strip allows the roof-rafter tails to slide past.

BUILD THE RIGHT SIDEWALL

Start by using a power miter saw to cut the 2×4 parts to length for the right-hand sidewall. (For this shed, the studs are 6 ft. 7¼ in. long.) Fasten the frame together using a cordless impact driver and 3-in. decking screws (1). Note that we framed two window openings into this wall, even though the wall has no windows. Adding the 24¼-in.-wide by 25¼-in.-tall rough openings now makes it much easier to install 24-in.-sq. barn-sash windows if desired in the future.

Before installing the plywood siding, check the wall frame for square by measuring its diagonal dimensions. Pull two tape measures across the wall frame, one from each diagonal corner (2). If the two measurements are identical, then the frame is square. If necessary, strike the corner of the frame with a hammer to knock it into square.

Cut ³⁄₈-in.-thick rough-sawn plywood siding 8½ in. longer than the height of the wall frame. Set the plywood siding into place on top of the wall frame (3). Align the plywood flush with the top of the wall, so that it extends 8½ in. past the bottom plate. Now, when the wall is tipped up into place, the extra plywood will overlap and hide the floor frame. Fasten the plywood to the wall frame with 1⁵⁄₈-in. decking screws.

Next, use a tablesaw to cut a 2-in.-wide starter strip from a pine 1×4 or 1×6. Start by tilting the sawblade to 45° and cutting a chamfer along the edge of the pine board. Then, adjust the sawblade to 90° and set the saw fence 2 in. from the blade. Now, rip the chamfered edge from the 1×4 to create a 2-in.-wide starter strip. (Chamfering the edge first is much safer than ripping the board to 2 in. and then trying to chamfer the narrow strip.)

Nail the strip to the top of the wall with its chamfered edge facing up (4). A starter strip is required at the top of each sidewall to provide

5 Nail batten strips to the wall, spacing them 12 in. on center and aligning each one with a 2×4 stud. Fasten the battens with 2½-in. nails.

6 Carefully tip the completed wall up into place, making sure it doesn't slide off the floor deck. If it's very windy, get two more helpers.

a surface for butting battens against. The chamfer is necessary to accommodate the rafter tails, which come down from the roof at a 45° angle and extend past the sidewalls.

Cut and nail batten strips to the sidewall (5). Then stand the wall up along the right side of the shed floor (6). Install two temporary 2×4 braces to securely hold the wall upright and perfectly plumb (7).

Secure the bottom of the wall by first driving 3-in.-long decking screws from the outside through the battens and plywood siding and into the floor frame (8). Drive one screw every 24 in. Then, move inside the shed and fasten the bottom wall plate to the floor with 4-in.-long washer-head structural screws (9). Space the screws about 16 in. apart and position them close to the outer edge of the 2×4 plate so they'll bite solidly into the rim joist, and not just into the plywood floor.

Screw a 2×4 brace to each wall end. Hold the wall plumb with a 4-ft. level, then screw the lower end of the brace to the floor frame.

Fasten the bottom of the wall with 3-in.-long screws spaced 24 in. apart. Drive the screws through the battens and into the floor frame.

Secure the bottom wall plate with 4-in.-long structural screws. Place the screws near the edge of the 2×4 so they'll go into the rim joist.

1 Use 1⅝-in. decking screws to attach rough-sawn plywood siding to the wall frame. Drive the screws slightly below the surface.

2 Drill a ¾-in.-dia. access hole through the plywood siding. Then cut out the window openings with a router and flush-cutting bit.

3 Set the cellular-PVC windowsill into the window's rough opening and then fasten it with three 2½-in.-long trim-head screws.

4 Install the head casing so that it extends past each side casing by an equal amount. Then attach the head casing with 2½-in. screws.

BUILD THE LEFT SIDEWALL

The left-hand sidewall is identical to the right-hand wall, except that it actually does have two 24-in.-sq. barn-sash windows. Frame each rough window opening to 24¼ in. wide by 25¼ in. high. Fasten together the 2×4 wall frame, check it for square, and then screw down the rough-sawn plywood siding **(1)**. Cut and install a chamfered starter strip along the top of the wall.

Rout the plywood from the window openings with a router and flush-cutting bit. Move the router in a clockwise direction to cut through the plywood and reveal the rough openings **(2)**.

The barn-sash windows purchased for this shed are made out of Azek, which is cellular PVC (polyvinyl chloride). Each sash comes with a trim kit that includes a windowsill, two side casings, and a head casing. Set the windowsill into the rough opening and screw it to the rough sill with 2½-in.-long trim-head screws **(3)**.

Set the 2-in.-wide batten strips into place on the sidewall. Space the strips 12 in. on center and align each one with a 2×4 wall stud.

Frame the rough opening for the doorway by screwing a horizontal 2×4 header to the jack studs (trimmers) on each side of the opening.

Cut the rough-sawn plywood siding to extend 8½ in. past the bottom wall plate. Attach the siding with 1⅝-in. screws.

Screw the PVC trim to the window opening, starting with the windowsill, followed by the two side casings, and then the head casing.

Attach the two vertical side casings, making sure each one extends into the rough opening by ¾ in. Set the horizontal head casing on top of the side casings and screw it to the wall **(4)**.

Complete the wall by cutting and attaching the batten strips **(5)**. Again, space the battens 12 in. on center and nail each one to a stud. Then cut short batten strips to fit snugly between the head casing and starter strip.

FRAME THE FRONT GABLE-END WALL

Cut the 2×4 parts for the front gable-end wall, which has a doorway and a window opening. Screw the frame together, using double studs at the sides of the doorway opening. The king studs run all the way from the bottom plate to the top plate, just like the regular wall studs. The jack studs, also called trimmers, extend from the bottom plate up to the horizontal header. Screw the jack studs to the king studs, then screw the header to the tops of the trimmers **(1)**.

Check the wall frame for square by measuring its diagonal dimensions. Tap the frame into square, if necessary, then install the rough-sawn plywood siding **(2)**. Use 2½-in. trim-head screws to fasten the PVC trim to the window's rough opening **(3)**.

Remove the 2×4 diagonal bracing from the right-hand sidewall. Tilt up the front gable-end wall and slide it tightly against the right-hand sidewall **(4)**. Secure the two walls by screwing the two corner studs together **(5)** (see p. 84).

Fasten the bottom of the wall to the floor frame by driving 3-in.-long decking screws through the battens and plywood siding. Go inside the shed and drive 4-in.-long washer-head structural screws down through the bottom wall plate and into the rim joist **(6)** (see p. 84).

Tip up the front gable-end wall, being careful it doesn't slide off the floor. Then, push it tight against the right-hand sidewall.

TOOL TIP

We chose to screw together the wall frame, but it would've been faster and cheaper to use a pneumatic framing nailer. The benefit of screws is that they can easily be removed if you need to disassemble the frame to correct a mistake or fix a problem, such as a warped or cracked stud.

Fasten the front wall to the sidewall by driving 3-in. screws through the corner studs. Space the screws about 16 in. apart.

Secure the bottom of the wall to the floor by driving 4-in.-long structural screws through the bottom plate and into the rim joist below.

INSTALL THE REAR GABLE-END WALL

Build the frame for the rear gable-end wall, which includes a rough opening for a 10-in.-high by 6-ft.-long transom window. Screw the rough-sawn plywood to the wall frame, then rout out the window opening. Set the PVC transom window into the rough opening and secure it with 2½-in. trim-head screws (1).

Next, nail on the batten strips, spacing them 12 in. on center (2). Note that the transom window is flush with the top of the wall, so there are no battens above the window. Now move the completed rear wall off to one side.

After routing the opening in the plywood siding, install the transom window. Fasten the sash to the rough opening with 2½-in. screws.

Nail the 2-in.-wide pine batten strips to the wall, spacing them 12 in. on center. Be sure to center each batten over a 2×4 wall stud.

Raise the left sidewall and slide it up against the front wall. Then, from outside, push the bottom of the wall tight to the floor frame.

Use 1⅝-in. screws to fasten the plywood siding at the ends of the walls. Corner boards will eventually conceal these screw heads.

Tilt up the left-hand sidewall and butt it tightly against the front gable-end wall (3). Screw together the two corner studs to tie the left sidewall to the front wall. Go around to the other side and use 3-in. screws to fasten the bottom of the wall to the floor frame. Then use 1⅝-in. decking screws to attach the plywood siding at the end of the walls to the corner studs (4). Go back inside the shed and screw through the bottom wall plate with 4-in.-long washer-head structural screws.

Now, set the rear gable-end wall in between the two sidewalls (5). Fasten this wall just as you did the other three walls: at the corners (inside and out), along the exterior bottom through the battens, and down through the bottom wall plate.

HELPFUL HINT

Transom windows are ideal for storage sheds because they let in a surprising amount of sunlight but take up very little wall space, which can be used for storage racks, shelving, or a workbench.

Set the rear gable-end wall into position between the two sidewalls. Be sure it's securely screwed in place before letting it go.

Securely fasten one resting cleat to each corner of the shed. Use three 3-in.-long screws and drive each one solidly into a wall stud.

Lift up the gable-end truss and set it down onto the resting cleats, which should be screwed in place about 3 ft. above the ground.

Roof Framing

With the walls up you can now frame the roof, which goes surprisingly fast because you'll be using the trusses built earlier. The standard roof trusses are lightweight and easily handled by a single person. However, you'll need at least three, or better yet, four able bodies to lift the gable-end trusses into place. Fortunately, there's a little trick that you can employ to make this job easier.

RAISE THE GABLE-END TRUSSES

Start by installing the two gable-end trusses, one at each end of the building. However, to lift the trusses over 8 ft. in the air without a crane is virtually impossible, so here's the trick: Make a pair of plywood resting cleats and "walk" the truss up the wall. Each cleat consists of a 12-in. by 16-in. piece of ½-in. plywood and a 12-in.-long 2×3; refer to the drawing on the facing page for more details.

Fasten one cleat to each wall corner with three 3-in. screws **(1)**. Position the cleats about 3 ft. off the ground, and be sure to drive the screws into a wall stud. Now lift the gable-end truss and set it onto the resting cleats **(2)**.

Next, lift the left end of the truss about 2 ft. above the resting cleat. Unscrew the cleat, move it up 2 ft., and screw it back to the wall corner. Lower the left end of the truss, allowing it to rest on the newly positioned cleat.

Repeat these steps for the right end of the truss: Lift it 2 ft. above the cleat, move up the cleat, and set the truss down onto the cleat. Continue in this manner, alternately raising each end of the truss and repositioning the cleats until you get within 12 in. or so of the top of

HELPFUL HINT

If you can't install the roofing for a few days, check the weather forecast. If rain is predicted, cover the entire plywood roof deck with plastic tarps. Then tie or screw down the tarps to prevent them from being blown off.

Raise one end of the truss about 2 ft., then move up the cleat. Repeat for the other end of the truss, slowly "walking" the truss up the wall.

When the truss gets to within 12 in. or 16 in. of the top of the wall, lift it off the resting cleats and set it down on top of the 2×4 wall plate.

the wall (3). As you're "walking" the truss up the wall, be sure to have someone on a ladder inside the shed steadying the truss to keep it from flopping over (4).

To raise the truss to its final resting spot will take at least three people, but again, four is better. Lift the truss off the resting cleats and set it on top of the wall. Push in on the bottom of the truss to make sure it's tight against the wall below. Then, drive 3-in.-long washer-head structural screws down through the 2×4 mounting plate and into the top wall plate (5).

Secure the truss to the top of the gable-end wall by driving 3-in. screws down into the top wall plate; space the screws 12 in. apart.

RESTING CLEAT

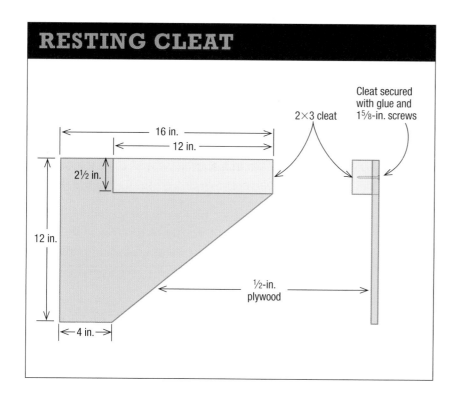

16 in.

12 in.

2½ in.

2×3 cleat

Cleat secured with glue and 1⅝-in. screws

12 in.

½-in. plywood

4 in.

Set a standard roof truss into place atop the sidewalls. Position one truss every 2 ft. and set each one directly over a 2×4 wall stud.

Fasten the trusses by driving a 3-in. decking screw up through the top wall plate and into the bottom edge of the 2×4 chord.

After installing all the trusses, cut a subfascia from a perfectly straight 2×4 to span across the rafter tails at each sidewall.

Securely screw the subfascia to the rafters by driving two 3-in.-long galvanized decking screws into the end of each rafter tail.

Repeat the previous five steps to install the remaining gable-end truss onto the opposite end of the shed. Use the resting cleats to carefully raise the truss up the wall, and be sure that someone keeps hold of the truss from inside the shed.

INSTALL THE STANDARD TRUSSES

The four standard roof trusses are installed along the tops of the sidewalls in between the two gable-end trusses. They're spaced 24 in. apart and each sits directly over a pair of wall studs. That way, the weight of the roof is properly supported and transferred from the rafters, down through the studs, to the floor frame and onto the foundation blocks.

Carry one standard truss through the doorway and into the shed. Raise the truss and set it on top of the sidewalls (1). Position the truss 24 in. from the gable-end truss and align each end of its 2×4 horizontal chord directly over a wall stud. Secure the truss to each sidewall by driving a 3-in.-long decking screw up through the top wall plate and into the chord.

Continue to install the remaining three trusses in a similar manner. Be sure to set each truss directly over a pair of wall studs and secure the ends with 3-in. screws (2).

Once the trusses are installed, cut a 2×4 subfascia to span across the rafter tails at each sidewall (3). Fasten the subfascia to the rafter tails with 3-in. decking screws (4). Install the second 2×4 subfascia to the rafter tails along the opposite sidewall.

5 Nail ½-in. plywood sheathing to both sides of the roof frame. Fasten the sheathing to the rafters with 2½-in. nails spaced 10 in. apart.

6 Cut a 4¼-in.-wide pine fascia board to span across the roof. Nail the fascia to the 2×4 subfascia with 2½-in. nails spaced 12 in. apart.

Next, cover the exposed roof frame with ½-in. CDX plywood. Fasten the plywood sheathing to the rafters with 2½-in. nails spaced 10 in. to 12 in. apart (**5**). Be sure all vertical end joints fall along the center of a rafter. And cut the plywood about 1½ in. short at the peak on both sides of the roof to create a gap for a ridge vent.

Cut a 4¼-in.-wide fascia board from a pine 1×6 and nail it to the subfascia (**6**). Be sure the top edge of the fascia is flush with the top of the 2×4 subfascia. Then use a handsaw to trim the rake boards flush with the fascia (**7**).

Before proceeding any further, apply a coat of paint or stain to the board-and-batten siding, rake boards, and fascias. Here, we applied solid-color stain using a 2-in. sash brush and ⅜-in.-nap paint roller (**8**).

7 Trim the rake boards flush with the fascias. At the end of each cut, reduce the downward pressure on the saw to avoid splitting the rake.

Roofing

As mentioned in the chapter introduction, the roof is covered with a unique faux-slate roofing product from DaVinci Roofscapes called Multi-Width Slate Tiles. Made of a super-durable plastic composite, the slate tiles (shingles) are 18 in. long and come bundled in an assortment of five different widths ranging from 6 in. to 12 in. They can be cut with a utility knife, circular saw, or power miter saw.

We installed the tiles in a 1-in. staggered pattern, but they also can be nailed in a straight line, as is common with most roof shingles.

8 Use a paint roller to apply paint or stain to the board-and-batten siding. Use a 2-in. sash brush to spread finish onto the batten edges.

Cover the plywood roof deck with 30-lb. felt. Position the first felt piece at the bottom of the roof, flush with the drip-edge flashing.

Install a second piece of felt underlayment to the center portion of the roof, making sure it overlaps the first piece by at least 3 in.

PREP THE ROOF DECK

Start by nailing aluminum drip-edge flashing to the edge of the roof, including along the eaves and up the rakes. Next, cover the plywood roof deck with 30-lb. felt underlayment, which is also called builder's paper. Roll out the first 36-in.-wide piece of felt along the eave **(1)**.

Align the felt even with the aluminum flashing, smooth out any wrinkles and trapped air, and then fasten to the plywood roof deck with 1-in. roofing nails. Install the next piece of felt, making sure to overlap the first piece by at least 3 in. **(2)**.

INSTALL THE STARTER COURSE

The very first row of roofing is called the starter course. It provides support for the first course of roofing and fills the spaces between the first-course tiles so you don't see the roof deck.

Begin by snapping a chalkline across the roof, positioned 11 1/2 in. up from the aluminum flashing **(1)**. The chalkline will provide a straightedge guide for nailing on the starter course of slate tiles, which are 12 in. long. By setting the starter tiles on the chalkline, they'll extend 1/2 in. past the aluminum flashing, creating the appropriate drip edge.

Hold the first starter tile in position with its upper end on the chalkline and its left edge extending 1/2 in. past the rake. Nail the slate tile to the roof with two 1 1/2-in.-long hot-dipped galvanized roofing nails **(2)**.

Set the next tile 3/8 in. away from the first tile. This space is necessary to provide room for the tiles to expand when the sun warms them.

TOOL TIP

If you're using a utility knife to cut the faux-slate tiles, consider upgrading to carbide-tipped blades. They cost a bit more than standard utility knife blades, but they cut cleaner and stay sharper longer.

Before nailing down the starter course, measure up 11½ in. from the drip-edge flashing and snap a chalkline across the roof.

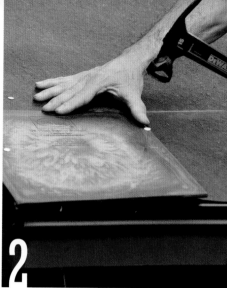

Set the first starter tile into place with its upper edge on the chalkline and its left edge protruding exactly ½ in. past the rake board.

Continue to nail starter tiles across the roof, aligning each one with the chalkline. And don't forget to leave a ⅜-in. expansion gap between the tiles (3).

At the end of the starter course, you'll likely have to cut the last tile to fit. When you do, be sure that it extends ½ in. past the rake.

BEGIN ROOFING

Once the starter course is installed you can nail down the first course of slate tiles. But before doing so, remember these three important rules:

1. First-course tiles must be aligned flush with the starter tiles along the drip edge and at the rakes.

2. Leave a ⅜-in. expansion gap between the tiles in the first course.

3. The gaps between the first-course tiles and starter tiles should be offset by a minimum of 1½ in.

For more information, refer to the Composite Slate Roof drawing on p. 92.

Start the first course by laying a slate tile on top of the starter course. Hold the tile flush with the starter course and fasten it with two 1½-in.-long hot-dipped galvanized roofing nails. Continue to nail slate tiles along the first course. Mix and match various-width tiles to ensure each tile overlaps the gap in the starter course below by at least 1½ in. (1) (see p. 92).

Continue to nail starter tiles across the roof, setting each one on the chalkline and leaving a ⅜-in. expansion space between the slate tiles.

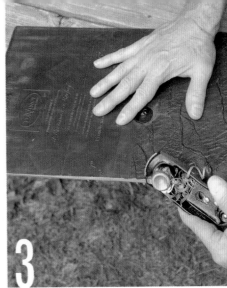

1 **Install first-course tiles** across the roof, making sure each one overlaps the expansion gaps in the course below by at least 1½ in.

2 **Cut the slate tiles** with a sliding compound miter saw. Install with the cut edge facing the adjacent tile, not the end of the roof.

3 **Use a block plane** to round over and soften the sharp, freshly cut edge. A sanding block with 80-grit sandpaper could be used instead.

COMPOSITE SLATE ROOF

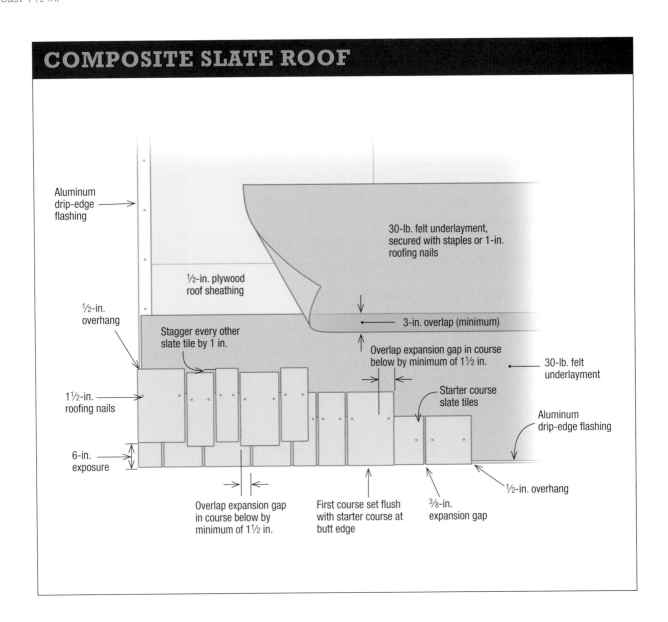

Aluminum drip-edge flashing

½-in. plywood roof sheathing

30-lb. felt underlayment, secured with staples or 1-in. roofing nails

3-in. overlap (minimum)

½-in. overhang

Stagger every other slate tile by 1 in.

Overlap expansion gap in course below by minimum of 1½ in.

30-lb. felt underlayment

1½-in. roofing nails

Starter course slate tiles

Aluminum drip-edge flashing

6-in. exposure

½-in. overhang

Overlap expansion gap in course below by minimum of 1½ in.

First course set flush with starter course at butt edge

⅜-in. expansion gap

When installing the second-course tiles, nail one tile with a 6-in. exposure, then set the next tile 1 in. lower to create a staggered pattern.

To ensure the roofing courses remain straight and level, measure up from the drip edge and snap chalklines at 6-in. intervals.

At the end of the roof you may need to cut the last tile to fit. The easiest way to cut the resilient composite tiles is with a sliding compound miter saw (2). However, the saw can't quite cut all the way through the 18-in.-long tile. So, after the initial cut, flip over the tile and complete the cut. Then use a block plane to round over the sharp edge left by the sawblade (3).

After completing the first course, begin the second course of tiles. Snap a chalkline across the roof 6 in. above the first-course tiles. Nail the first tile in the second course on the chalkline to produce a 6-in. exposure to the weather. Hold the second tile in place, ⅜ in. away from the first tile, but drop it down 1 in. to create the staggered pattern (4). Continue installing slate tiles in this manner across the roof, staggering every other tile by 1 in. And remember to leave ⅜-in. gaps between the tiles and to offset the joints from one course to the next by 1½ in.

To make sure the courses continue to run perfectly straight, measure up from the drip-edge flashing and snap chalklines onto the roof deck every 6 in. (5). Then set the upper ends of the slate tiles that are exposed 6 in. to the weather on the chalklines. Stagger every other tile 1 in. lower.

HELPFUL HINT

Installing the slate tiles in a staggered pattern produces a much more interesting-looking roof, but it does take quite a bit longer than running the roofing courses in straight lines. And the staggered pattern requires more attention to avoid mistakes, such as misaligned tiles and incorrect weather exposure.

To make it easier to reach the upper half of the roof, install a pair of metal roof brackets. Then nail a staging plank to the brackets.

As you work your way up the roof, keep ⅜-in. gaps between the slate tiles and maintain the 1-in. stagger along the butt edges.

FINISH ROOFING

Once you've installed slate tiles about halfway up the roof, stop and install a pair of roof brackets and a staging plank. That way, you'll be able to work safely and more comfortably on the upper half of the roof. Securely nail each metal bracket to a roof rafter, not just to the plywood sheathing. Then set the plank in place and nail it to each bracket **(1)**.

Cover the exposed plywood on the upper half of the roof with 30-lb. felt, tacking down the felt with 1-in. roofing nails and overlapping the piece below by at least 3 in. Continue installing slate tiles, one course at a time all the way up the roof **(2)**.

At the roof peak, nail down full-length slate tiles and allow them to run long past the peak. Secure each tile with two 1½-in.-long nails.

Use a cordless circular saw to trim the slate tiles flush with the edge of the plywood roof deck, which is 1½ in. down from the peak.

5 Adhere a wide strip of granulated self-sealing ice-and-water barrier to the top surface of the rigid-polypropylene ridge vent.

6 Press the ridge vent down onto the roof peak, and then secure it with 2½-in.-long roofing nails spaced about 12 in. apart.

When you get to the last course, don't bother cutting each slate tile—just let them run long past the peak (**3**). Once the last course is installed, use a cordless circular saw to cut all the slate tiles in a single pass (**4**). Now move around to the other side of the shed and install slate tiles to the opposite roof plane.

The final roofing step is to install a ridge vent along the peak of the roof. The ridge vent works in conjunction with the soffit vents (to be installed next) to exhaust hot, stale air from inside the shed. Here, we installed a low-profile polypropylene ridge vent manufactured by Cor-A-Vent.®

Begin by covering the top surface of the 11½-in.-wide ridge vent with a layer of granulated self-sealing ice-and-water barrier. This step isn't mandatory, but it does add an extra level of water- and weather-resistance to the roof peak. Peel off the protective backing from the ice-and-water barrier to expose its sticky adhesive surface. Then press the barrier down onto the ridge vent (**5**). Set the vent onto the roof peak, making sure it overlaps each roof plane by an equal amount. Fasten the ridge vent with 2½-in.-long roofing nails (**6**).

Now cover the ridge vent with 6-in.-wide ridge tiles, exposed 6 in. to the weather (**7**). Secure each ridge tile with two 2½-in. roofing nails. Place the nails about 7 in. up from the butt edge so the next tile will cover them.

7 Cover the ridge vent with 6-in.-wide ridge tiles, which are available from the roofing manufacturer. Fasten each tile with two nails.

Install a perforated vinyl soffit vent underneath the overhanging eave of the roof that runs along the top of each sidewall.

INSTALL SOFFIT VENTS

Soffit vents allow fresh air to enter the shed under the eaves and exhaust out the ridge vent. Soffit vents also help keep out bugs, birds, and other critters. Here, we installed a 6¾-in.-wide perforated vinyl soffit vent, but aluminum vents are also available. Cut the vent to length with aviation snips to fit under the eaves along each sidewall **(1)**. Slip the vent into place and secure it to the underside of each rafter tail with 1¼-in. decking screws **(2)**. Repeat to install a soffit vent under the opposite eave.

Hold the soffit vent against the underside of the overhanging eave and then fasten it to each rafter tail with 1¼-in. decking screws.

HELPFUL HINT

If you can only find 12-in.-wide perforated soffit vents, cut them to width by slicing them in half lengthwise with a sharp utility knife.

To form a corner board, attach a 4-in.-wide pine board to the edge of a 3-in.-wide pine board using 2½-in.-long galvanized nails.

Angle the corner board into place, tucking its upper end tightly against the overhanging face of the plywood siding on the gable-end wall.

ATTACH CORNER BOARDS

Each exterior corner of the shed is protected by a vertical trim piece that's appropriately called a corner board. Preassemble each corner board by nailing a 4-in.-wide pine board to a 3-in.-wide pine board (1).

Next, cut a 6-in.-wide strip of 30-lb. felt and staple it to the inside of the L-shaped corner-board assembly. If water should penetrate the corner board, the felt will block it from seeping into the corner of the shed. Press the corner-board assembly tightly against the shed corner (2). Fasten it with 2½-in. galvanized nails spaced 16 in. apart (3). Install the remaining three corner boards in a similar manner.

Nail the preassembled corner board to the shed corner with 2½-in. nails. Later, stain or paint the corner board to match the siding.

When making the door-assembly jig, use a framing square to ensure that the 2×4 cleat is perfectly square with the long 2×2 fence.

Make the door panel out of five 1×6 tongue-and-groove pine boards. Set the boards into the jig one at a time and tap them together.

Cross-Buck Batten Door

To complement the shed's board-and-batten siding, we built a traditional cross-buck batten door. The door measures 34 in. wide by 74 in. tall and is composed of 14 pine boards: five tongue-and-groove 1×6s, which form the door panel; a perimeter frame consisting of three 6-in.-wide horizontal rails and four 2-in.-wide vertical stiles; and two 6-in.-wide diagonal cross-buck battens.

Once the ¾-in.-thick pine boards are cut to size, the door can be assembled in about an hour or so. However, to ensure the door is perfectly flat, square, and straight, it's important to build it on top of a large, flat, stable workbench.

CONSTRUCT THE DOOR

Start by cutting five tongue-and-groove 1×6s to 74 in. long for the door panel. Then cut three 6-in.-wide by 34-in.-long rails and four 2-in.-wide by 28-in.-long stiles. Finally, cut two 6-in.-wide by 43-in.-long cross-buck battens.

Next, to ensure the door comes out perfectly square, build a simple door-assembly jig. Screw to the workbench an 8-ft.-long fence made from a straight 2×2 or 2×4. Cut a 36-in.-long 2×4 cleat and set it at a right angle to the fence. Use a framing square to make sure the cleat is perfectly square with the fence, then screw the cleat to the workbench **(1)**.

Now, prep the door panels by first using a tablesaw to rip the groove off one 1×6. Then rip the tongue from a second 1×6. Take the first 1×6, the one without the groove, and place it in the jig with its square-cut edge against the fence. Butt its end against the 2×4 cleat.

CROSS-BUCK BATTEN DOOR

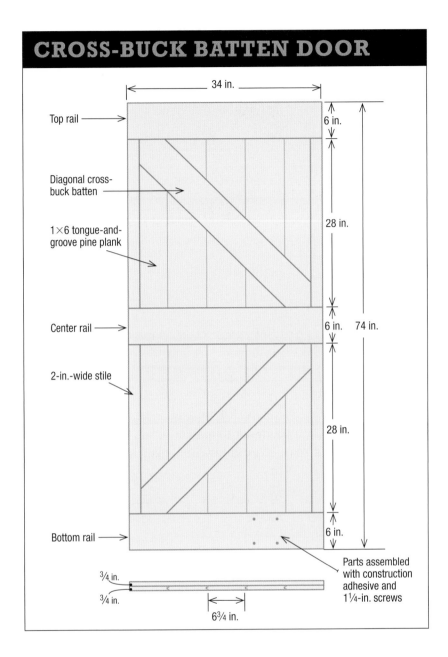

Top rail

Diagonal cross-buck batten

1×6 tongue-and-groove pine plank

Center rail

2-in.-wide stile

Bottom rail

34 in.

6 in.

28 in.

6 in.   74 in.

28 in.

6 in.

Parts assembled with construction adhesive and 1¼-in. screws

¾ in.

¾ in.

6¾ in.

Cross-Buck Backward?

Cross-buck batten doors are often installed with the flat door panel facing in toward the shed and the cross-buck battens facing out. That seems to make sense since the stile-and-rail frame and diagonal battens are attractive and visually interesting. But if that's true, why did we install the door with the flat panel facing out? Here's why: The door sheds rain and snow much better with the flat panel exposed to the weather. When the stile-and-rail frame faces out, water and snow dams up along the horizontal rails and in the corners of the diagonal battens. And that can lead to water damage, blistered paint, and eventually rot.

If you decide to install the door with the stile-and-rail frame facing out, here's one way to mitigate—though not totally eliminate—water damage: Bevel the top edge of the horizontal rails to help drain off rain and melting snow.

Take one of the unaltered 1×6s and set its grooved edge over the tongue protruding from the first board. Push the boards tightly together to close the tongue-and-groove joint. Slide the third board into position, align its groove over the tongue on the second board, and push the joint closed (2). If necessary, use a wood block and hammer to tap the joints closed. Repeat to install the fourth 1×6.

Complete the door panel by installing the 1×6 that had its tongue trimmed off earlier. Slide the board's groove over the tongue on the fourth board and tap the joint closed. Before proceeding any further, check to be sure that all the tongue-and-groove joints are closed and the end of each 1×6 is pressed tightly against the 2×4 cleat.

Apply a bead of construction adhesive across the door panel, then press the 6-in.-wide horizontal rail down into the adhesive.

Hold the 2-in.-wide stile flush with the outer edge of the door panel. Fasten the stile to the door with four 1¼-in.-long decking screws.

Next, apply a zigzag bead of construction adhesive across the door panel. Press the 6-in.-wide bottom rail into the adhesive. Align the rail flush with the bottom of the door panel and fasten it with 1¼-in. decking screws (**3**). Drive four screws into each 1×6 board.

Apply a narrow bead of construction adhesive near both edges of the door panel. Set the 2-in.-wide stiles into place and fasten them with 1¼-in. screws (**4**). Continue in this manner to attach the center rail, the final two stiles, and the top rail. In each instance, fasten the parts with adhesive and 1¼-in. screws.

FIT THE DIAGONAL BATTENS

Cutting the diagonal cross-buck battens to fit precisely within the door's stile-and-rail frame is no easy feat. Fortunately, we developed a

MARKING JIG

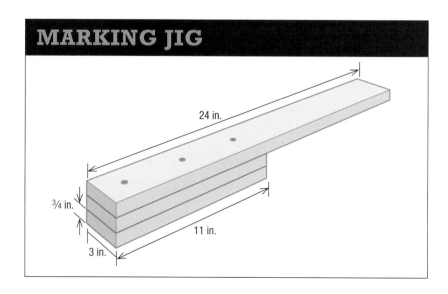

24 in.

¾ in.

11 in.

3 in.

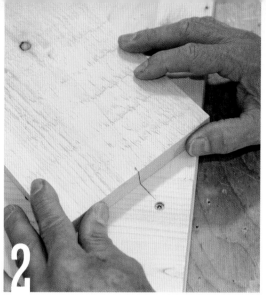

Lay a long straightedge rule diagonally across the door and mark the position of each inside corner onto the stile-and-rail frame.

Set the cross-buck batten into place on the door panel. Align the centerline mark on the batten with the lines drawn onto the frame.

Use a homemade marking jig to quickly and accurately locate and draw the angled cutlines onto the 6-in.-wide cross-buck batten.

shop-made marking jig and measuring technique that greatly simplifies the job and produces perfect-fit parts.

Make the marking jig from a 48-in.-long pine 1×4. Rip the 1×4 to 3 in. wide, then crosscut it to produce two 11-in.-long base parts and one 24-in.-long scribing arm. Stack and nail together the two shorter parts to create a 1½-in.-thick by 3-in.-wide by 11-in.-long base. Then, nail the 24-in.-long scribing arm to the base; see the Marking Jig drawing on the facing page.

Next, lay a long steel rule (or other straightedge) diagonally across the stile-and-rail frame on the door. Align each end of the rule precisely with the inside corner of the frame. Draw pencil lines along the rule and onto the frame (1).

Mark a centerline onto each end of the 6-in.-wide cross-buck batten. Lay the batten diagonally across the door with the pencil mark on each end aligned with the ones drawn onto the frame (2).

Now, set the marking jig onto the door panel with the scribing arm extending over the diagonal cross-buck batten. Hold the jig's base against the 6-in.-wide door rail, and draw a cutline along the scribing arm and onto the cross-buck batten (3). Lift the jig, rotate it 90°, and set it down onto the door panel. Hold the jig against the 2-in.-wide stile and draw a second cutline onto the batten. Repeat to mark cutlines onto the opposite end of the batten. Bring the batten to the power miter saw, align the blade with the angled cutlines, and trim both ends of the batten (4).

Apply a zigzag bead of construction adhesive diagonally across the door panel. Set the cross-buck batten into place and secure it with 1¼-in. screws, driving four screws into each 1×6 board (5).

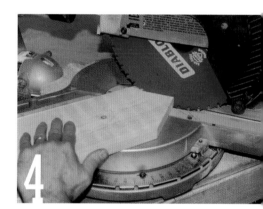

Adjust the miter saw's blade to match the angle of the cutlines. Then cut an angled point onto each end of the cross-buck batten.

Apply a zigzag bead of construction adhesive diagonally across the door panel and push the cross-buck batten into the adhesive. Fasten it with 1¼-in.-long decking screws.

1 Screw the side casing to the right-hand side of the doorway. Locate the pine casing ¾ in. away from the edge of the doorway opening.

2 To create a clearance gap between the door and side casing, tap nails partway into the siding immediately adjacent to the casing.

3 Hold the door in place over the doorway opening. Slide it up against the spacer nails, and then screw the hinges to the side casing.

4 Attach the side casing to the left side of the doorway and the head casing on top of the side casings. Paint or stain the door and casings, as desired.

Repeat the previous steps to mark and cut the remaining cross-buck batten. Then, mount three strap hinges to the right side of the door.

HANG THE DOOR

The shed door is mounted to the door casing on the right side of the doorway. Start by cutting two side casings and one head casing from 1-in.-thick rough-sawn pine; make each piece 3 in. wide.

Fasten one side casing to the right-hand side of the doorway opening. Position the casing ¾ in. away from the edge of the opening and use a 4-ft. level to make sure the casing is perfectly plumb. Secure the casing with 3-in. screws (1).

Next, hold a 2½-in. nail against the edge of the casing and hammer it partway into the siding (2). Position this nail about 8 in. up from the shed floor. Tap in another nail about 8 in. down from the top of the doorway. These two nails will act as temporary spacers, creating the correct gap between the door and casing.

Set the door in place and hold it tightly against the spacer nails. Have a helper screw the hinges to the side casing (3). Pull out the spacer nails and check to make sure the door swings open and closed smoothly.

Screw the remaining side casing to the left-hand side of the doorway. Position it ⅛ in. away from the door edge. Set the head casing on top of the two side casings and fasten with 3-in. screws (4). Install a handle and latch to the door.

Use the power miter saw to cut one end of each 4×4 rafter to 20°; square-cut the other end of each 13-in.-long rafter to 90°.

Adjust the miter-saw blade to 45° and cut a shallow chamfer into all four corners on the square-cut end of each 4×4 rafter.

Door Overhang

A door overhang is a short roof that extends out over an exterior door. It's an architectural element that's seldom seen on sheds, but we added one here because it's attractive and functional: The overhang diverts rain and snow away from the door. Plus, it gave us an excuse to install some more faux-slate roof tiles.

The door overhang could be built and attached to the shed one board at a time, but it's easier to prefabricate it in the shop and then install it as a fully assembled unit. Then, only the roofing must be added after installation.

MAKE THE SUPPORT STRUCTURE

The roof overhang is supported by two diagonal 4×4 braces and two 4×4 rafters. The rafters are topped with 1½-in.-thick by 5-in.-wide tongue-and-groove pine planks, which are then covered with faux-slate roofing.

Start by using the miter saw to cut two 13-in.-long rafters. Square-cut one end of each 4×4 rafter, and then miter-cut the opposite end to 20° (1). Now set the saw's miter angle to 45° and chamfer all four corners of the square end of each rafter (2).

Next, use the miter saw to cut two 19-in.-long diagonal braces. Miter-cut one end to 60° and trim the other end to 10°. Now, fasten the rafters to the diagonal braces with 10-in.-long structural screws (3).

Drill ³⁄₁₆-in.-dia. pilot holes through the rafters, and use 10-in.-long structural screws to fasten the rafters to the diagonal braces.

After bevel-ripping the groove off the first pine plank, screw the plank to the 4×4 rafter. Be sure the plank extends 1 in. past the rafter.

Set the groove in the second pine plank over the tongue on the first plank. Tap the joint closed. Fasten the plank with 3-in. screws.

Use the tablesaw to rip the tongue off the third plank. Set the plank into position on the roof overhang, then screw it to the 4×4 rafters.

ATTACH THE ROOF DECK

Use the miter saw to cut three 1 1/2-in.-thick by 5-in.-wide tongue-and-groove pine planks to 52 in. long. Now move to the tablesaw and tilt the sawblade to 20°. Bevel-rip the grooved edge off one of the pine planks.

Hold the beveled plank against the top of the rafter with its end extending 1 in. past the side of the rafter. Fasten the plank with two 3-in.-long screws **(1)**. Repeat to attach the opposite end of the plank to the other rafter.

Install the next plank, fitting its groove over the protruding tongue of the previous plank **(2)**. Be sure the ends of the two planks are flush, then screw the second plank to the rafters. Use the tablesaw, with its blade set at 90°, to rip the tongue off the final pine plank, making it 4⅝ in. wide. Attach the plank to the rafters with 3-in. screws **(3)**.

INSTALL THE OVERHANG

Hold the door overhang in place above the doorway with its pine-plank roof deck about 2 in. below the lower edge of the upper gable-end wall. Mark pencil lines along the bottom of the diagonal braces and onto the siding. Take down the door overhang and set it off to one side. Measure 2½ in. up from each pencil line and drill a ⅛-in.-dia. "witness" hole through the siding.

Screw a 2×4 between the wall studs on each side of the doorway. The 2×4s provide solid blocking for attaching the roof overhang.

Secure the bottom end of the diagonal braces by driving 3-in. screws through the braces and into the 2×4 blocking behind.

Go inside the shed and attach a horizontal 2×4 between the studs on each side of the doorway (1). Position each 2×4 directly over the ⅛-in. witness holes. Now set the door overhang back into place above the door. Drive one 3-in. decking screw at an angle through the roof deck and into the king stud on each side of the doorway.

Fasten the bottom end of each diagonal brace with two 3-in. screws (2). Be sure the screws go through the brace and into the horizontal 2×4s installed earlier.

Cover the roof overhang with 30-lb. felt. Next, install a starter course of slate tiles, followed by multi-width slate tiles arranged in a staggered pattern (3). Be sure to tuck the last course of tiles underneath the siding on the upper gable-end wall.

Nail faux-slate tiles to the roof deck of the door overhang. Stagger the butt joints on the tiles to match the pattern on the shed roof.

DESIGN DETAILS

1

2

1. The shed's roof is covered with plastic composite tiles that look like real slate. The tiles were installed in a staggered pattern.

2. The windows and trim are made of cellular PVC, a resilient, water-resistant material that won't rot, split, or ever need painting.

3. The rear gable-end wall has a 20-in.-sq. diamond-shaped window installed high above a narrow 6-ft.-long transom window.

4. The 34-in.-wide pine door is protected by a door overhang, and the window to the right is one of five windows in this compact shed.

5. The shed's left-hand sidewall has two barn-sash windows and board-and-batten siding finished in barn-red solid-color stain.

3

4

5

4

VINYL-SIDED
STORAGE SHED

When we first conceived this attractive, neatly trimmed shed, we focused on three design goals: simple construction, spacious storage, and a virtually maintenance-free exterior. We're glad to report that we met all three goals and couldn't be happier with the results.

At 10 ft. by 16 ft., the shed offers plenty of storage space for a lawnmower, wheelbarrow, garden tools, bicycles, and many other items that typically clog up a garage. It sits on a solid-concrete block foundation, which installs quickly with very little excavation.

The floor frame is built of 2×6s; the walls and roof are framed with 2×4s. The walls are sheathed in oriented-strand board (OSB), and the roof rafters are covered with exterior-grade plywood and architectural-style roof shingles. A pair of sliding doors glides open to reveal a nearly 5-ft.-wide doorway.

To create a low-maintenance, long-lasting exterior, the shed features ivory-colored vinyl siding, white PVC trim boards, vinyl soffit vents, vinyl gable-end vents, and a white vinyl-clad double-hung window. These surfaces will never need painting, scraping, or sanding, and they won't ever rot, crack, or suffer water damage. Only the wood doors will occasionally require some attention to maintain their painted exteriors. (To order building plans for the Vinyl-Sided Storage Shed, see Resources on p. 214.)

Foundation and Floor Frame

The foundation consists of 4-in.-thick by 8-in.-wide by 16-in.-long solid-concrete blocks arranged in three parallel rows. Because the building site wasn't level, we stacked blocks on the low end to raise it level with the high end.

SET THE FOUNDATION BLOCKS

Begin by laying out 12 solid-concrete foundation blocks in three parallel rows. Make each row 15 ft. 6 in. long, and space the first and third row 9 ft. 10 in. apart, as measured from outer edge to outer edge. Center the middle row of blocks in between the two outer rows. (For more details, see the Foundation-Block Layout drawing on p. 112.)

Next, dig 2 in. to 3 in. of soil out from beneath each block. Fill the hole with gravel and compact the surface with a hand tamper or 4×4 post. Replace the blocks and then use a 9-in. torpedo level to ensure each block is level in two directions: along its length and across its width.

Now lay a long, straight 2×4 or 2×6 on top of one row of blocks. Set a 4-ft. level on top of the board and check to make sure that the row is level. If the ground is sloping, shim up the low blocks with pieces of solid, weather-resistant material, such as additional 2-in.- or 4-in.-thick solid-concrete blocks, pressure-treated wood, composite lumber, or asphalt roof shingles (1).

Stack 4-in.-thick concrete foundation blocks to build up the low end of the site. Set 2-in.-thick patio blocks on top, if necessary.

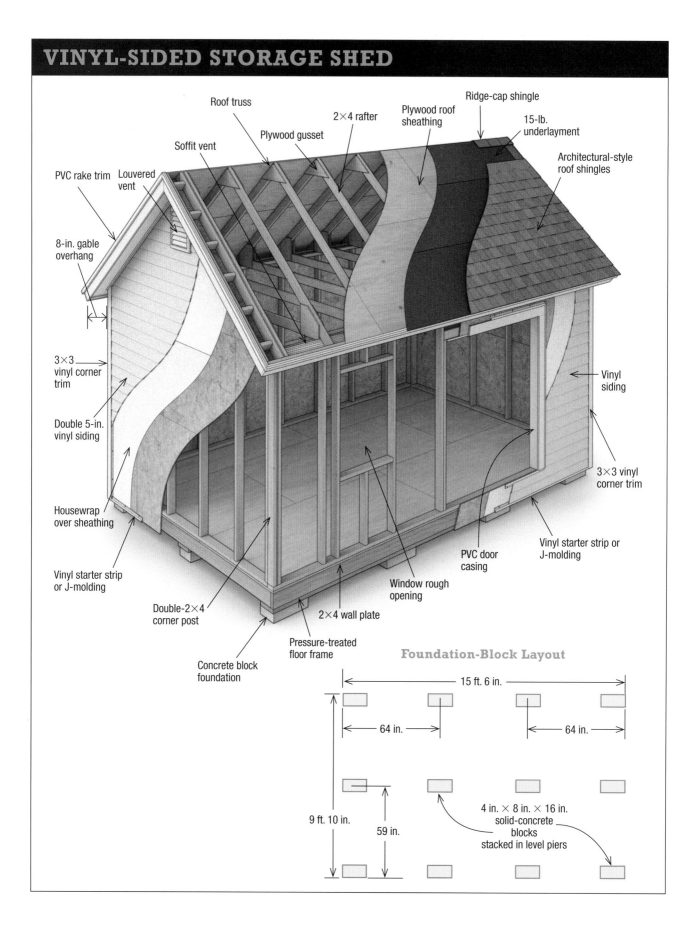

Roof truss

2×4 rafter

Plywood roof sheathing

Ridge-cap shingle

15-lb. underlayment

Plywood gusset

Soffit vent

Architectural-style roof shingles

PVC rake trim

Louvered vent

8-in. gable overhang

3×3 vinyl corner trim

Double 5-in. vinyl siding

Vinyl siding

3×3 vinyl corner trim

Housewrap over sheathing

Vinyl starter strip or J-molding

Vinyl starter strip or J-molding

PVC door casing

Double-2×4 corner post

Window rough opening

2×4 wall plate

Pressure-treated floor frame

Concrete block foundation

Foundation-Block Layout

15 ft. 6 in.

64 in.

64 in.

9 ft. 10 in.

59 in.

4 in. × 8 in. × 16 in. solid-concrete blocks stacked in level piers

Set the L-shaped assembly on the foundation blocks with the 2×8 mudsill against the blocks and the 2×6 rim joist sticking straight up.

Install the 2×6 floor joists, spacing them 16 in. on-center. Use 3-in. screws to fasten the joists at each end and in the middle.

FRAME THE FLOOR

Start by cutting three pressure-treated 2×8 mudsills to 15 ft. 8 in. long; these boards will sit directly on the foundation blocks. Next, cut two 2×6 rim joists to the same length: 15 ft. 8 in. Fasten each 2×6 rim joist to a 2×8 mudsill with 3-in. galvanized decking screws, driving the screws through the sill and into the edge of the rim joist. Set the L-shaped assemblies on top of the outer two rows of foundation blocks, with the 2×8 mudsill down flat against the blocks (1). Lay the remaining 2×8 mudsill across the row of blocks in the center of the foundation.

Next, cut 13 floor joists from pressure-treated 2×6s. Make each joist 9 ft. 9 in. long. Screw the first and last floor joists between the rim joists. Check the frame for square by measuring the diagonals; if necessary, adjust the frame until the two dimensions are equal. Now install the remaining 2×6 floor joists, spacing them 16 in. on center (2).

INSTALL THE FLOOR DECK

Cover the floor frame with ¾-in. exterior-grade or pressure-treated plywood (1). Keep the plywood flush with the perimeter edges of the floor frame, and be sure to align all end-butt joints over the center of a floor joist. Secure the plywood to the joists below with 2-in. galvanized decking screws (2). Space the screws 10 in. to 12 in. apart and drive them slightly below the surface of the plywood.

Cover the shed's floor frame with ¾-in.-thick weather-resistant plywood. Align all end-butt joints over the center of a floor joist.

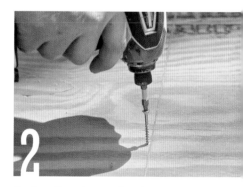

Fasten the plywood to the joists with 2-in. decking screws. Here, we pulled a taut nylon line to indicate the center of the floor joist.

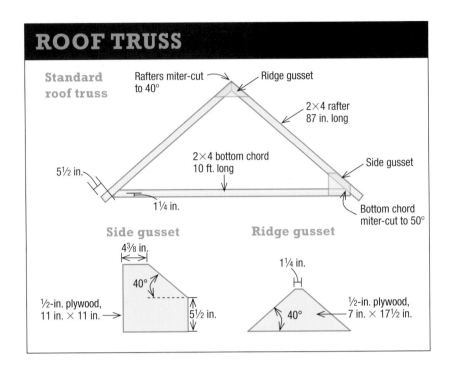

Roof Trusses

The roof frame consists of seven standard trusses spaced 24 in. on center and two gable-end trusses. These last two trusses are installed atop the end walls of the shed and feature fly rafters that extend out past the gable-end walls.

Each truss is built of two 2×4 angled roof rafters and a horizontal 2×4 chord. The parts are held together with ½-in. plywood gusset plates that are glued and nailed across the joints.

BUILD STANDARD ROOF TRUSSES

For each of the nine trusses, use a power miter saw to cut two 2×4 rafters to 87 in. long. Miter-cut the upper end of each rafter to 40° (for a 10-in-12 roof pitch). Then, cut a 2×4 bottom chord for each truss to 120 in. long, mitering both ends to 50°. (For more details, see the Roof Truss drawing above.)

Next, saw gusset plates from ½-in. exterior-grade plywood. Cut triangular ridge gussets to 7 in. high by 17½ in. wide. Cut the side gussets to 11 in. square, then saw off one corner, as shown in the Roof Truss drawing. You'll need a total of 16 ridge gussets: two for each standard truss, one for each gable-end truss. And you'll need 36 side gussets, four per truss. Also, cut eight 4-in. by 6-in. plywood stop blocks.

To build a roof truss, set the two 2×4 rafters and 2×4 bottom chord into place atop the template fastened to the plywood floor deck.

Apply a generous bead of construction adhesive across the joint where the two rafters meet at the peak, and then fasten the 1/2-in. plywood gusset plate to the rafters with 1 1/2-in.-long roofing nails.

Lay out one roof truss on the plywood floor deck, then temporarily screw the boards to the deck. Next, screw two plywood stop blocks to the outer edges of the bottom chord and two stop blocks to the outer edges of each rafter. Finally fasten a stop block to the end of each rafter tail.

Use this template to assemble the seven standard roof trusses. Set one pair of 2×4 rafters and one bottom chord on top of the template (1). Fasten the boards together with gusset plates, construction adhesive, and 1 1/2-in. roofing nails (2).

Flip the truss over and attach gusset plates to the other side. Repeat the previous steps to build six more standard trusses. Fabricate one more truss, but nail gusset plates to only one side. This truss will become one of the two gable-end trusses. Unscrew the stop blocks and fasten gusset plates to one side of the truss that's screwed to the floor deck. This final truss will become the second gable-end truss.

BUILD GABLE-END ROOF TRUSSES

Each gable-end truss requires two additional 2×4 structural supports. One is a collar tie installed between the rafters 19 in. above the bottom chord, which adds rigidity to the gable-end truss and provides a solid surface for nailing on sheathing. Fasten the collar tie to the truss with two side gussets. The other 2×4 support is an 8-ft.-long shoe plate that gets fastened flush with the bottom chord. When the truss is installed, the shoe plate sits on the top plate of the gable-end wall.

Sheathe the outer surfaces of the two gable-end trusses with 7/16-in. OSB. Fasten the sheathing with 2-in. screws.

After assembling the gale-end trusses, sheathe their exterior surfaces with 7/16-in.-thick OSB (1).

Build the fly rafters by setting 5-in.-tall blocks of 2×4s in between two 2×4 roof rafters. Fasten with 3½-in.-long nails, driving two nails through the rafters and into the end of each 2×4 block.

Sheathe the underside of the fly rafters with strips of 7/16-in. OSB. Fasten the sheathing with 2-in. galvanized screws or nails.

HELPFUL HINT

Rub wax or bar soap onto screw threads and the screws will spin into the wood much faster and easier.

FORM THE GABLE OVERHANG

The gable overhang on each of the gable-end trusses is essentially an 8-in.-deep eave framed with 2×4s. For each gable-end truss, cut four 87-in.-long 2×4 rafters and 12 short 2×4 blocks to 5 in. long. Stand six blocks between two rafters (spaced about 15 in. apart) and fasten them with 3½-in. nails to form a fly rafter **(1)**. Repeat to assemble the remaining two rafters and six blocks. Use 3-in. decking screws to attach the fly rafters to the top edges of the truss.

Cut 8-in.-wide strips of OSB to fit along the underside of the fly rafters; fasten the strips with 2-in. nails or screws **(2)**. Repeat the previous steps to build a gable overhang onto the remaining gable-end truss.

Wall Framing

Each of the four shed walls was framed on the floor deck with 2×4s and 7/16-in. OSB sheathing before being raised. You can also attach the siding while the wall is on the deck, but vinyl siding goes up rather quickly and easily, so we waited until the walls were erected.

BUILD THE GABLE-END WALLS

Begin by building the two shorter gable-end walls. Each wall includes a top and bottom horizontal wall plate and six vertical wall studs spaced 24 in. on center.

For each wall, cut two plates to 9 ft. 5 in. long, and six 2×4 studs to 79¼ in. Fasten the studs between the plates with 3½-in. nails **(1)**.

Nail together the 2×4 frame for the gable-end wall. Space the studs 24 in. on-center and secure each end of every stud with two nails.

Cover the exterior of the gable-end wall with 7/16-in. OSB. Fasten the sheathing with 2-in. nails spaced 8 in. to 10 in. apart.

Cut the sheathing to protrude 8 in. beyond the bottom wall plate. This overhang is necessary to conceal the floor frame.

Then sheathe the exterior of the wall frame with 7/16-in. OSB. Hold the sheathing flush with the top wall plate, and secure it with 2-in. nails (2). Cut the OSB to extend 8 in. past the bottom plate (3). Later, when you stand up the wall, the overhanging plywood will conceal the floor frame. Repeat to build the opposite gable-end wall.

BUILD THE SIDEWALLS

Next, use 2×4s to construct the two long sidewalls. The rear sidewall is a simple frame; it has no windows or doors. The front sidewall has rough openings for the window and doors.

Build the rear sidewall first: Cut two wall plates to 15 ft. 8 in. long and 11 studs to 79¼ in. Also cut two 3½-in.-wide by 79¼-in.-long strips of ½-in. plywood. Set each plywood strip between two 2×4 studs and nail them together to form two corner posts. Nail the studs between the plates and position one corner post at each end of the wall frame (1). Cover the exterior of the wall frame with 7/16-in. OSB.

Now, build the front sidewall. Cut two wall plates to 15 ft. 8 in. long and 10 wall studs to 79¼ in. long. Cut two 3½-in.-wide strips of ½-in. plywood and make two corner posts, as for the rear wall.

Nail the corner posts and wall studs between the top and bottom wall plates, following the spacing shown in the Front Sidewall drawing on p. 118. Create the rough door opening with two 70½-in.-tall jack studs and a 63-in.-long header. Set the header between the king studs and then nail through the king studs and into each end of the header (2). Locate the rough window opening 30 in. from the left-hand corner post.

Fasten a corner post at each end of the sidewalls. Make the post by sandwiching a ½-in.-thick plywood strip between two 2×4s.

Slip the header between two jack studs that form the door's rough opening. Push the header down tight against the top of the jack studs, and then drive nails through the king studs and into the ends of the header.

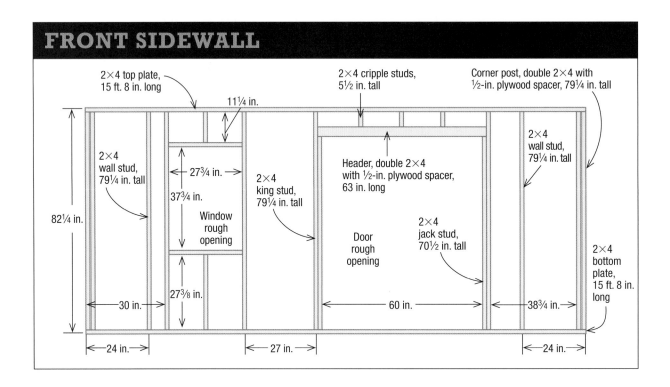

2×4 top plate, 15 ft. 8 in. long

11¼ in.

2×4 cripple studs, 5½ in. tall

Corner post, double 2×4 with ½-in. plywood spacer, 79¼ in. tall

2×4 wall stud, 79¼ in. tall

82¼ in.

27¾ in.

37¾ in.

Window rough opening

2×4 king stud, 79¼ in. tall

Header, double 2×4 with ½-in. plywood spacer, 63 in. long

Door rough opening

2×4 jack stud, 70½ in. tall

2×4 wall stud, 79¼ in. tall

2×4 bottom plate, 15 ft. 8 in. long

27⅜ in.

30 in.

24 in.

27 in.

60 in.

38¾ in.

24 in.

ERECT THE WALLS

Lay the rear sidewall on the floor deck and then raise it into position (1). Fasten the wall by nailing or screwing down through the bottom plate (2). Repeat to raise the front sidewall.

Now install each of the two end walls. Set the bottom plate on the floor deck and tip the wall into place (3). Fasten the bottom plate to the floor deck, then screw the end stud to the corner post on the adjacent sidewall (4). Once you've erected all four walls, screw the bottom edge of the OSB sheathing to the floor frame.

Lift up the rear wall, being careful it doesn't slide off the floor deck. Then attach two 2×4 braces to temporarily hold the wall upright.

Secure the wall by nailing down through the bottom plate and into the floor frame. Drive three or four nails into each stud bay.

Tip the end wall into position between the two side-walls. If necessary, push out the top of each sidewall to squeeze in the end wall.

Fasten together each wall corner by driving 3-in. screws through the end stud and into the corner post. Space the screws 10 in. to 12 in. apart.

Roof Framing

With the trusses already built, framing the shed's roof goes pretty quickly. It took three of us about an hour to set all nine trusses. Two people can easily install the seven standard trusses, but you'll need at least three people to raise the much-heavier gable-end trusses.

SET THE GABLE-END TRUSSES

Set one of the gable-end trusses on top of the sidewalls, so that it's hanging upside down. Carefully rotate the truss up to its final resting spot atop the end wall (1). Secure the truss by driving 3-in. screws down through the 2×4 shoe plate and into the top wall plate (2). Repeat the previous steps to install the second gable-end truss onto the opposite end wall.

While standing on sturdy stepladders, slowly tip the gable-end truss up into position. Be very careful not to push it past vertical.

Securely fasten the truss to the end wall by driving 3-in. screws down through the 2×4 shoe plate and into the top wall plate.

Lift the first standard truss into place and set it on top of the sidewalls. Align each end of the truss directly over a 2×4 wall stud.

Drive a 3-in. screw up through the top wall plate and into the bottom chord on the truss. Drive in another screw on the other side of the stud.

SET THE STANDARD TRUSSES

The seven standard trusses are spaced 24 in. on center, but don't bother measuring. Simply set each truss directly over a wall stud. This not only speeds up the installation, but more importantly it transfers the weight of the roof to the wall frame and down to the floor frame and foundation blocks.

Stand the first truss on top of the sidewalls directly over a stud **(1)**. Hold the truss steady as a helper drives a 3-in. screw at an angle up through the top wall plate and into the bottom chord **(2)**. Drive in a second screw from the opposite side of the stud. Then screw down the other end of the truss. Continue in this manner to set the remaining trusses.

Nail a 2×4 subfascia across the rafter tails. It's important to hold the subfascia perfectly flush with the top edges of the roof rafters.

Cover the roof frame with ½-in. plywood sheathing. Fasten the plywood with 2-in. nails and center all end-butt joints over a rafter.

Create a maintenance-free exterior with rake boards cut from PVC trim. Nail the 5½-in.-wide plastic trim to the gable overhangs.

Nail a PVC fascia to the subfascia along the edges of the roof. The white plastic trim doesn't require painting and it'll never rot.

Next, nail a 2×4 subfascia across the rafter tails on each side of the roof (**3**). Cover the roof frame with ½-in. exterior-grade plywood (**4**).

INSTALL THE PVC ROOF TRIM

The last step before nailing on the roof shingles is to trim out the shed's roof with ¾-in.-thick PVC boards. Start by installing 5½-in.-wide PVC trim along the rake and fascia. Use the miter saw to cut the upper end of each rake board to 40°. Then nail the rake to the outer edge of the gable overhang (**1**). Cut the fascia from the same 5½-in.-wide PVC board. Hold the end of the fascia flush with the rake board, then nail it to the 2×4 subfascia (**2**). To add more depth and shadow lines to the shed roof, nail 1½-in.-wide PVC trim to the rake boards and fascia (**3**).

Use 2½-in. nails to attach 1½-in.-wide PVC trim to the rake boards and fascia. Hold the trim flush with the plywood roof sheathing.

Staple 15-lb. underlayment to the plywood roof. Start at the eave and work up toward the ridge. Overlap each course by at least 4 in.

Roofing

The roof for this 10-ft. by 16-ft. shed required three squares (300 sq. ft.) of architectural-style shingles. You'll also need about 70 lin. ft. of starter shingles and 18 lin. ft. of ridge shingles. Note that the following step-by-step sequence is based on the installation of these particular roof shingles, which measure 13¼ in. wide by 39⅜ in. long. If the shingles you're installing are larger or smaller, you'll need to adjust the shingling pattern and spacing.

INSTALL THE STARTER COURSE

Start by covering the plywood roof sheathing with 15-lb. asphalt-saturated underlayment, also called builder's paper. Secure the underlayment with ⅜-in. staples **(1)**. Then screw a 1-in.-thick wood strip to the rake trim on each end of the roof **(2)**. These temporary strips act as alignment guides for installing the shingles: Cut the shingles flush with the strips to create the proper overhang.

HELPFUL HINT

Don't install asphalt roof shingles on an extremely hot day. Walking on the sun-softened shingles will wear away the surface granules and shorten the life of the roof.

Screw a temporary alignment strip to the rake boards. Cut the strip from scrap wood that's ¾ in. to 1 in. thick, but no thicker.

Install a course of starter shingles along the lower edge of the roof. Secure each shingle with four 1¼-in.-long roofing nails.

Now fasten starter shingles along the lower edge of the roof (**3**). Hold the shingles flush with the PVC fascia trim and secure each one with four 1¼-in. roofing nails. At the end of the roof, allow the starter course to overlap the temporary alignment guide. Use a utility knife to trim the shingle flush with the guide.

Next, install starter shingles up each end of the roof (**4**). Keep the shingles flush with the alignment guide. Repeat to install starter shingles to the opposite side of the roof.

TOOL TIP

A sharp utility knife is adequate for cutting asphalt roof shingles, but you can speed up the job by renting a manual shingle cutter. The tool looks somewhat like a giant paper cutter and easily slices through the toughest shingles.

Continue the starter course of shingles up the sloped ends of the roof. Fasten each shingle with four 1¼-in.-long roofing nails.

Snap shingle layout lines across the roof with a chalk reel. Space the lines 5⅝ in. apart to create the proper exposure to the weather.

Start the first course with a full-length shingle. Hold it flush with the starter course and secure it with four 1¼-in. roofing nails.

SHINGLE THE ROOF

Snap chalklines across the roof to serve as layout lines for the shingle courses. Snap the first line 13 in. up from the lower edge of the starter course. Then snap a line every 5⅝ in. all the way up the roof **(1)**.

Starting at one end of the roof, nail down the first roof shingle flush with the starter course **(2)**. Next, cut 6½ in. off the left end of a shingle and use the remainder to start the second course. (Trimming the shingle automatically staggers the vertical seams between courses by

Cut and install the first shingle in the next two courses. Note that it's necessary to trim each shingle to produce staggered end joints.

Continue installing roof shingles until you've covered approximately half of the roof. Be sure each course is flush with a chalkline.

To make it safer and easier to shingle the upper roof half, work from staging planks supported by metal roof brackets.

6½ in.) Nail the first shingle in the second course flush with the chalk-line to create a 5⅝-in. exposure to the weather. Now cut 13 in. off a roof shingle and use it to start the third course (**3**). Repeat this pattern as you make your way up the roof: Trim 19½ in. off a shingle to begin the fourth course, 26 in. off to start the fifth course, and 32½ in. to start the sixth course. Save the cut off pieces of shingles; you can use them later. When you reach the seventh course, start with a full-length shingle and begin the pattern all over again.

Continue shingling until you've covered about half of the roof (**4**). Then stop and install four roof brackets and two 8-ft.-long staging planks. Secure each bracket with three 2½-in. nails, and be sure to drive the nails into a rafter.

Cover the upper half of the roof with asphalt-saturated underlayment (**5**). Resume shingling until you reach the ridge. Now remove the roof brackets and staging planks and shingle the other side of the roof.

Once you've shingled both roof planes, cover the peak with ridge-cap shingles. Fasten each ridge shingle with two 2-in. roofing nails (**6**).

Protect the ridge of the roof with ridge-cap shingles. Center the shingles on the peak and fasten each one with two 2-in. roofing nails.

1 Cover the shed walls with housewrap. Staple the wrap to the sheathing, then slice the wrap from the vent, window, and door openings.

2 Place the mounting base of the louvered vent into the hole at the top of the gable-end wall. Fasten the PVC base with roofing nails.

Vinyl Siding

The shed walls are covered with 12-ft.-long panels of double 5-in. vinyl siding. Each panel is 10 in. wide, but molded into two 5-in. courses, which have a realistic wood-grain texture that's reminiscent of cedar bevel siding. Vinyl siding cuts easily and goes up quickly, but installation can't begin until you prep the walls and trim out the shed.

PREP THE SHED FOR SIDING

Cover the shed walls with housewrap, which acts as an air-infiltration barrier and also protects the OSB sheathing should any moisture get trapped behind the siding. Fasten the housewrap to the sheathing with 3/8-in. staples **(1)**.

HELPFUL HINT

If the window doesn't sit plumb and level in the opening, have a helper slip shims beneath the window from inside the shed while you hold the window from outside and check it with a level.

TOOL TIP

A pneumatic nail gun provides the quickest, easiest way to fasten PVC trim boards, but you can also use a hammer and galvanized finishing nails. Just be sure to bore nail-pilot holes first.

To help seal out weather, line the window opening with self-adhesive flashing. Apply the waterproof flashing to the sill, then each side, and finally across the top of the window opening.

Secure the window with 1 1/4-in. roofing nails. Drive one nail through each prepunched hole in the window's mounting flange.

Cut a PVC head jamb to fit across the top of the doorway opening. Fasten it to the header with a pneumatic nailer and 2½-in.-long nails.

Cut a PVC side jamb to fit along each vertical side of the doorway. Securely nail each jamb, spacing the nails 10 in. to 12 in. apart. Nail the PVC side casings over the side jamb.

Cut a PVC head casing to span the top of the doorway. Set it on top of the two vertical side casings, then secure it with 2½-in. nails.

Use a utility knife to cut the housewrap from the rough openings for the door, window, and two louvered vents. Each PVC vent consists of a mounting base and a snap-on louvered faceplate. Set the mounting base in the vent opening at the top of the gable-end wall (2). Secure it with 1¼-in.-long roofing nails. Don't snap on the louvered faceplate until after the vinyl siding has been installed (see p. 131).

Next, apply self-adhesive waterproof flashing around the perimeter of the window's rough opening. Stick the 4-in.-wide flashing along the sill first, and then apply a vertical strip to each side of the opening (3). Finish by applying flashing across the top of the window, making sure it overlaps the vertical side strips.

Tip the window into the rough opening. Check it for plumb and level, then nail it to the wall (4). Seal the window perimeter with four more strips of self-adhesive waterproof flashing.

TRIM THE WINDOW AND DOOR OPENING

Begin by crosscutting 5½-in.-wide PVC trim to 59½ in., which is the width of the door opening. Use a tablesaw to rip the board to 4½ in. wide. Hold the board—called the head jamb—against the top of the door opening and nail it to the header (1). Next, cut two 5½-in.-wide PVC trim boards to 71¾ in., then rip them to 4½ in. Nail these boards—called the side jambs—to each vertical side of the door opening (2).

Now cut three door casings from a 5½-in.-wide PVC board. Cut two side casings to 71¾ in. long and then rip them to 4 in. Nail one side casing to each side of the doorway. Cut the head casing to 62¾ in.,

Nail 1½-in.-wide PVC strips around the window. Note that the top and bottom strips extend past the two vertical side strips.

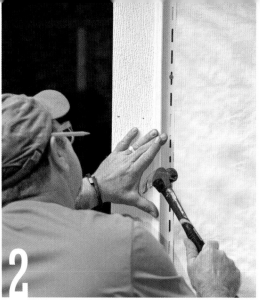

Nail the J-molding atop the head casing over the doorway. Leave the nail heads protruding a little to allow for expansion and contraction.

Install J-molding along each side of the doorway. The channel in the molding will conceal the ends of the vinyl siding panels.

Miter-cut J-molding to fit around the perimeter of the window. Press the molding tight to the PVC trim, then secure it with roofing nails.

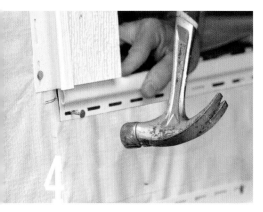

Cut a length of vinyl utility trim to fit below the door opening. Secure with roofing nails.

then rip it to 4 in. Nail the head casing across the top of the side casings **(3)** (see p. 127).

Next, use the tablesaw to rip 1½-in.-wide strips of PVC trim, and nail the strips around the perimeter of the window **(4)** (see p. 127).

INSTALL THE J-MOLDING

J-molding is used to conceal the ends of the vinyl-siding panels. Cut the vinyl molding and siding with a miter saw equipped with a high-speed-steel plywood blade. However, to produce the smoothest cuts, install the blade backwards so that its teeth are pointing up.

Rotate the sawblade to 45° and miter-cut a length of J-molding to fit across the top of the doorway. Set the J-molding on top of the head casing and secure it with 1¼-in. roofing nails **(1)**. Cut and nail J-molding to each side of the doorway **(2)**. Repeat to install J-molding around the perimeter of the window **(3)**.

Use aviation snips to cut a length of vinyl utility trim to fit beneath the doorway. Secure the 1½-in.-wide trim with roofing nails **(4)**. (Note that utility trim is also called undersill trim and is designed to accept siding panels that have had the nailing hems cut off their upper edges.) Then install utility trim along the very tops of the sidewalls, directly below the overhanging rafter tails.

INSTALL THE CORNER TRIM

To conceal the ends of the siding panels at each corner of the shed, install 3-in. by 3-in. vinyl corner trim. Cut the trim to length on the miter

HELPFUL HINT

Most PVC trim boards give you a choice of two finished surfaces: wood-grain texture on one side, smooth finish on the other. The wood-grain pattern is attractive—and it matches the siding—but the smooth side collects less dirt and dust, and is easier to scrub clean.

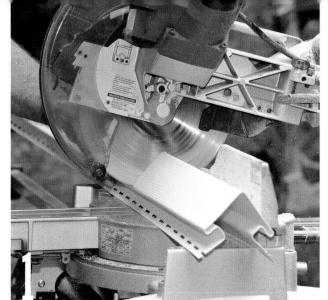

Cut 3-in. by 3-in.-vinyl corner trim to length on the miter saw. The backward-facing plywood blade cuts cleanly without chipping.

Hold the trim in place against the corner of the shed and then fasten with roofing nails driven into the center of the slots.

saw (1). Fasten the trim to the shed corner with 1¼-in. roofing nails spaced 10 in. to 12 in. apart along each of the two nailing flanges (2).

HANG SIDING ON THE GABLE-END WALL

The verb "to hang" is used to describe vinyl siding installation. That's because the siding panels "hang" off of nails. The nails aren't hammered tight against the siding. This is important, because vinyl siding must be free to expand and contract during temperature changes. For this same reason, it's necessary to cut the panels ¼ in. to ⅜ in. short at each end, so there's room for expansion.

Start by nailing a vinyl starter strip or J-molding along the bottom of the shed wall (1). Use a sliding-compound miter saw to cut a vinyl siding panel to fit between the two corner trim pieces. Again, be sure to cut the panel ¼ in. to ⅜ in. short at each end. Now bow the panel out slightly in the center and tuck each end behind the corner trim. Slide the panel down into the J-molding, then nail it to the wall (2).

HELPFUL HINT

The friction of sawing through vinyl occasionally leaves a rough burr along the just-cut edge. Allow the piece to cool for a few seconds, then use your thumb to rub off the burrs.

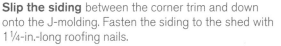

Nail a vinyl J-molding or starter strip along the bottom of the wall. Use a level to ensure that the molding or strip is perfectly level.

Slip the siding between the corner trim and down onto the J-molding. Fasten the siding to the shed with 1¼-in.-long roofing nails.

3

Slip the second siding course behind the corner trim, then pull up to securely lock its lower edge to the upper edge of the first course.

4

Bow out the center of each siding panel, then slip its ends behind each corner trim. If the panel fits too tightly, trim it a little shorter.

Cut to length a panel for the second course. Bow the panel out slightly and slip its ends behind the corner trim. Now snap its lower edge onto the receiving lip running along the upper edge of the first siding course (**3**). Press the lower edge of the second course to make sure it's fully locked in place along its entire length. Drive roofing nails through the center of the nailing slots to secure the second course. Leave each nail head protruding $1/32$ in. (or about the thickness of a dime).

Continue to cut and install vinyl-siding panels as you make your way up the wall, bowing out each panel and slipping it behind the corner trim pieces (**4**). Check to be sure the panels fit comfortably with at least $1/4$ in. of space at each end.

Nail J-molding to the underside of the gable over-hang. The $5/8$-in.-deep channel in the molding accepts the ends of the siding panels.

HELPFUL HINT

Fill all nail holes in the PVC trim with white silicone caulk. The caulk will seal out water and help keep the nails from rusting and staining the white trim.

5

Miter-cut the ends of the siding panel to 40° to fit the upper half of the gable-end wall. Be sure it fits loosely between the J-moldings.

Miter-cut siding pieces to fit along each side of the vent's base plate. Cut the siding about ¼ in. short at each end to allow for expansion.

Install J-molding along the underside of the gable overhang **(5)**. Then miter-cut the ends of the siding panels to 40° to fit between the J-molding strips **(6)**. Notch the siding to fit around the louvered vent.

Next, miter-cut progressively shorter pieces of siding to fit between the vent and the J-molding **(7)**. Cut a triangular piece of siding to fit at the very top of the wall. Squeeze a generous bead of construction adhesive into the channel running along the bottom edge of the siding piece, then press it onto the receiving lip of the previous siding course **(8)**. Complete the gable-end wall by snapping the vent's louvered faceplate onto the mounting base **(9)**.

Apply a thick bead of construction adhesive into the channel on the rear of the triangular piece of siding that fits at the peak. Push up on the triangular piece and lock it onto the receiving lip of the previous course.

Snap the louvered faceplate onto the vent's mounting base. Then check that the faceplate covers the ends of the adjacent siding panels.

Install a length of vinyl starter strip or J-molding perfectly level and flush with the very bottom edge of the front sidewall.

Mark the first course of siding where it overlaps the door casing. Cut the siding panel ¼ in. to ⅜ in. short to allow for expansion.

HANG SIDING ON THE SIDEWALL

Hanging siding on the sidewalls is similar to siding the gable-end walls. Start by nailing a starter strip or J-molding along the very bottom of the wall (1). Set the first siding panel into place and mark where it intersects the door casing (2). Notch the siding with aviation snips and a utility knife. Install the notched panel, then check to be sure it fits loosely at four locations: corner trim at left end, starter strip or J-molding along bottom edge, J-molding at the door casing, and utility trim beneath the doorway (3).

Install the notched siding panel by slipping it behind the corner trim and into the utility trim running beneath the doorway opening.

Cut siding panels to fit around the window and door. Be sure that the panels to the left of the door are level with the ones on the right.

Since the first siding course isn't long enough to span the entire sidewall, cut another piece of siding to reach the opposite corner trim. Overlap the two panels by 1 in. Continue to hang siding panels up the sidewall, cutting them to fit around the window and doorway **(4)**. Use the tablesaw to rip to width the uppermost siding course. Slip the siding into the utility trim fastened at the top of the wall **(5)**. Repeat the previous steps to hang siding on the remaining two walls of the shed.

Rip the last siding course to width, then slip its upper edge into the narrow utility trim running along the very top of the sidewall.

Cover the underside of the overhanging eave along each sidewall with a vinyl or aluminum soffit vent. If necessary, cut the vent to width.

Use ⁵⁄₈-in. screws to fasten the soffit vent to the 2×4 subfascia and to the rafter tails. The vent will admit fresh air but block out bugs.

INSTALL THE SOFFIT VENTS

Nail vinyl utility trim, also called undersill trim, to each side of the gable overhang and across the overhang at the peak. Secure the trim with 1¼-in. roofing nails.

Soffit vents allow fresh air to enter the shed where, through convection, it forces hot air out the gable-end vents. (The vents are also helpful in keeping out bugs and bees.) Soffit vents are simply long perforated panels made of vinyl or aluminum; here, we installed the vinyl type. Fasten the vents to the underside of the overhanging eaves at the top of the sidewalls. If you can't find 6-in.-wide soffit vents, buy 12-in. vents and slice them down the center.

Cut the soffit vent to length and press it up against the underside of the overhanging eave (1). Screw the lower edge of the vent to the 2×4 subfascia (2). Secure the vent's upper edge by driving one screw into each rafter tail.

TRIM THE GABLE OVERHANG

Insert the vinyl roll flashing into the utility trim, then slide it up toward the roof peak. Be sure the flashing fits loosely without binding.

There are a couple of ways to deal with the exposed OSB sheathing that's visible on the underside of the fly rafters. You could paint it or cover it with soffit vent or vinyl siding. Here, we concealed the OSB with vinyl roll flashing. The glossy-white flashing is easy to install, and it complements the white PVC trim.

Nail utility trim to each side of the gable overhang (1). Cut two short utility trim pieces to fit across the overhang at the peak. Also nail a short piece of utility trim at the lower end of each gable overhang. Now slip the vinyl flashing into the utility trim pieces (2). Pull the flashing all the way to the roof peak, then tuck its upper end into the utility trim nailed across the overhang. Trim the lower end of the flashing ¼ in. short of the fascia board, then slip its end into the utility trim nailed across the bottom of the overhang. Repeat to install vinyl flashing on the remaining gable overhangs.

Use a tablesaw to cut half-lap joints into the red-cedar door rails. Note that the rails were also rabbeted to receive the door panel.

Spread construction adhesive onto the stiles and press the rail into place. Repeat to install the rail on the opposite end of the doorframe.

Door Construction

We custom-built a pair of wood sliding doors, but you could also purchase wooden or steel doors. Note that we mounted the sliding-door hardware on the inside of the shed, not the outside. This protects the hardware from the weather, but it does create a small problem: You can't store items against the wall near the doorway because the doors need that space to slide open. If you'd rather not sacrifice that little bit of storage space, simply mount the doors on the outside of the shed.

ASSEMBLE THE DOORFRAME

The doorframe consists of six red-cedar parts: two vertical door stiles, two horizontal door rails, one vertical mid-stile, and one horizontal mid-rail. The door panel is cut from $7/16$-in.-thick exterior-grade plywood. (Refer to the Door Details drawing on p. 136.)

Start by cutting the rails and stiles to size. Then use the tablesaw to cut wide rabbets and half-lap joints into the top and bottom door rails (1). Also, cut $1/2$-in.-deep by $1\,3/4$-in.-wide rabbets into the right and left door stiles.

Apply construction adhesive to the half-lap joints and set the rails between the stiles (2). Clamp the four frame parts together, then measure the opposing diagonals to ensure the frame is square. Nail the half-lap joints with 1-in. brads. Leave the clamps in place for at least 1 hour.

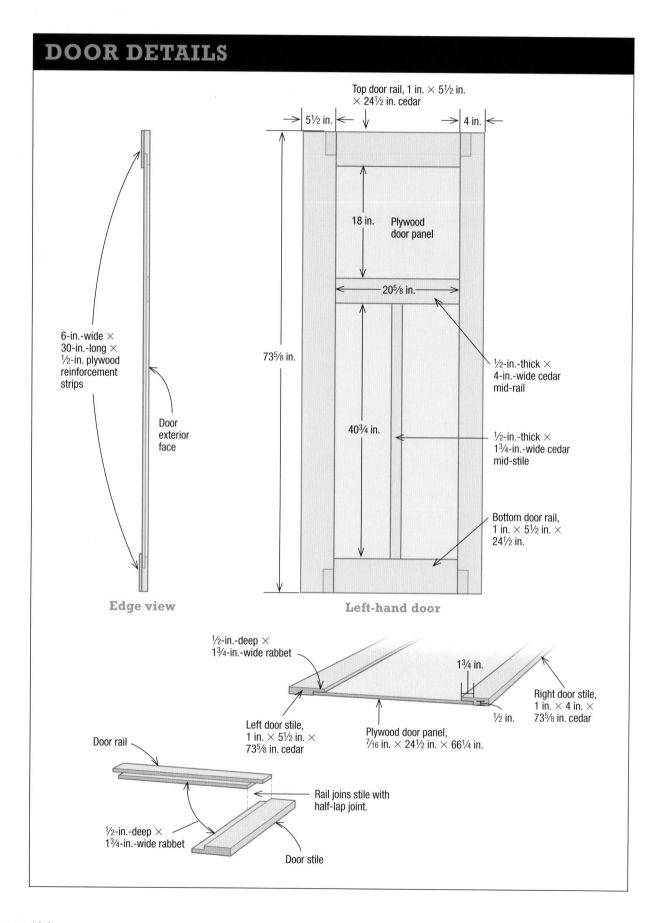

Top door rail, 1 in. × 5½ in. × 24½ in. cedar

5½ in.

4 in.

18 in.   Plywood door panel

20⅝ in.

73⅝ in.

40¾ in.

½-in.-thick × 4-in.-wide cedar mid-rail

½-in.-thick × 1¾-in.-wide cedar mid-stile

Bottom door rail, 1 in. × 5½ in. × 24½ in.

6-in.-wide × 30-in.-long × ½-in. plywood reinforcement strips

Door exterior face

Edge view

Left-hand door

½-in.-deep × 1¾-in.-wide rabbet

1¾ in.

½ in.

Left door stile, 1 in. × 5½ in. × 73⅝ in. cedar

Plywood door panel, 7/16 in. × 24½ in. × 66¼ in.

Right door stile, 1 in. × 4 in. × 73⅝ in. cedar

Door rail

Rail joins stile with half-lap joint.

½-in.-deep × 1¾-in.-wide rabbet

Door stile

1

2

3

Apply construction adhesive along the wide rabbet cut into the doorframe. Then set the plywood door panel down into the adhesive.

Fasten the 7/16-in.-thick plywood door panel with 3/4-in. screws. Drive the screws just slightly below the surface of the panel.

Glue and screw 6-in.-wide plywood reinforcement strips across the top and bottom of the door. The strips fortify the doorframe.

INSTALL THE DOOR PANEL

Run a thick bead of construction adhesive around the rabbet in the doorframe **(1)**. Press the plywood door panel into the rabbet, and secure it with 3/4-in. screws **(2)**. Next, glue and screw 6-in. by 30-in. plywood strips across the top and bottom of the door **(3)**. Flip the door over and install the 4-in.-wide horizontal mid-rail **(4)**. Then attach the 1 3/4-in.-wide mid-stile **(5)**. Repeat the previous steps to build the second door.

Cut a 4-in.-wide mid-rail from red cedar and set it between the door stiles. Adhere it to the plywood panel with glue and 1-in. brads.

Install the 1 3/4-in.-wide by 40-in.-long red-cedar mid-stile in the center of the door. Secure it with glue and 1-in. brads.

4

5

Align the center of the hangers 4 in. from the door edge, then drill clearance holes for carriage bolts. Tighten the three bolts with a wrench.

Secure the 10-ft.-long steel rail to the wall studs with lag screws. To ensure the doors hang properly, be certain the rail is perfectly level.

Insert the four-wheel truck into the steel rail. Then roll the door forward and guide the wheels on the second truck into the rail.

HANG THE DOORS

The sliding-door hardware consists of a 10-ft.-long steel rail and four steel door hangers, which are each equipped with a four-wheel truck. Attach two hangers to the top of each door (1). Then, use lag screws to fasten the steel rail to the wall studs inside of the shed (2). Stand one door near the end of the steel rail and guide the four-wheel truck into the rail (3). Lift the opposite side of the door and feed the second truck into the rail. Repeat for the second door. Adjust the hangers so that each door hangs perfectly plumb.

Entry Steps

Depending on how high the shed floor is above the ground, you may have to build steps or at least a small platform. The floor of our shed was about 24 in. high, so it was necessary to build a two-step platform. You can easily alter the following step-construction sequence to suit your needs.

Begin by using a shovel and garden rake to level the ground in front of the shed door. Then, form a solid base for the steps by spreading 2 in. to 3 in. of gravel onto the ground. Thoroughly compact the gravel with a hand tamper or 4×4 post.

Build a set of platform steps out of pressure-treated lumber. Cover the frame of each platform with 5/4 by 6-in. decking.

Build a rectangular frame for the first platform out of pressure-treated 2×6s or 2×8s. Make the frame 20 in. to 24 in. wide and 4 ft. longer than the width of the doorway. That way, it'll extend 2 ft. beyond each side of the door, providing a safe, comfortable way to enter and exit the shed. Assemble the four frame parts with 3-in. decking screws.

Now cut pressure-treated joists to fit within the frame. Screw the joists 16 in. on center, then cover the frame with pressure-treated 5/4 by 6-in. decking. Allow the decking to extend ¾ in. at the ends and along the front edge of the frame. Set the platform on the gravel bed, centered in front of the doorway.

Build the second platform the same as the first one, but make its frame 10 in. to 12 in. wide and 2 ft. longer than the doorway width. Install joists 16 in. on center, then set the frame on top of the first platform. Screw the second-platform frame to the first platform. Also, screw the frame to the underside of the shed's floor frame. Finally, cover the second platform with 5/4 by 6-in. decking (1).

HELPFUL HINT

Sliding doors are preferred over out-swinging doors in areas that receive a significant amount of snowfall. That's because if snow piles up against the shed, you can still slide open the doors. You wouldn't be able to open outswinging doors without first shoveling away the snow.

1.

2

1. The left gable-end wall has a louvered vent, which helps exhaust hot air. The ivory-colored vinyl siding spans between vinyl corner trim.

2. A vinyl-clad double-hung window on the front façade admits natural light and fresh air. The roof is covered with architectural-style shingles.

3. A pair of sliding doors permits easy access to the storage space. The 30-in.-wide doors roll on a steel track mounted inside the shed.

4. No windows were installed on the gable-end wall or rear wall so that there would be plenty of wall space for shelves and hanging hooks.

5. This shed features low-maintenance materials, including vinyl siding, 3-in.-wide vinyl corners, and PVC rake and fascia boards.

3

4 **5**

5

CEDAR-SHINGLE SHED

The design and construction of this quaint and quintessential New England storage shed is based on the weathered outbuildings and cottages that punctuate the shorelines of Connecticut, Rhode Island, and, in particular, Cape Cod, Massachusetts. And just like its coastal cousins, this 12-ft. by 16-ft. backyard beauty features white-cedar wall shingles, red-cedar roof shingles, painted frame-and-panel doors, bright-white trim, and double-hung windows adorned with flowerboxes.

Traditional architecture is reflected on the interior, as well: Exposed posts, beams, and diagonal bracing form a modified timber-frame building. And the walls and roof are sheathed with wide pine barn boards that are exposed on the interior, lending the look and feel of an old barn or dune shack.

To produce this traditional design we employed a modern framing method called timber-panel construction, which greatly simplified the wall framing. Each wall is composed of a rectangular 2×4 frame that supports a network of 4×4 posts, purlins, braces, and rough openings for windows and doors. The result is an easy-to-build storage shed that honors the traditional construction methods and designs of yesteryear. (For more information on the Cedar-Shingle Shed, see Resources on p. 214.)

Foundation and Floor Frame

The local building inspector required us to install a frost-proof foundation, so we dug holes down to the frost line and poured six 10-in.-dia. concrete piers. The shed's floor frame was then bolted to the piers.

If your town's building department approves an on-grade foundation, then you can build the shed on solid-concrete blocks, as shown for the Board-and-Batten Shed or Vinyl-Sided Storage Shed.

POUR THE CONCRETE PIERS

The six foundation piers are created by pouring concrete into 10-in.-dia. fiber-form tubes, which are often referred to by the trade name Sonotubes®. However, for each pier you must dig a 12-in.- to 14-in.-dia. hole down to the frost line. (If you don't know the frost-line depth in your area, ask the town's building inspector.) Making the holes a little wider than necessary provides extra room for shifting the fiber-form tubes precisely into position.

Dig the first two holes, then drop in the fiber-form tubes. Measure to confirm that the distance across the outer edges of the tubes is 12 ft. 2 in.

Lay out and mark the six holes in two parallel rows with three holes in each row. The two rows should be spaced 12 ft. 2 in. apart, as measured from the outside edge of one fiber-form tube to the outside edge of the other. And each row must be 16 ft. 2 in. long, again as measured from outside edge to outside edge. (For detailed information, refer to the Concrete-Pier Layout drawing on p. 148.)

Dig the first hole in each row down to the frost line. Set the fiber-form tubes into the holes, then measure across the tubes to confirm the outside-to-outside dimension of 12 ft. 2 in. (1). Dig the four remaining holes and install the fiber-form tubes. Measure the position of the tubes to make sure they're spaced properly. Check the tubes for plumb with a level, then backfill around each one with soil.

Next, use a laser level or line level and mason's string to strike a level line across all six fiber-form tubes. Adjust the height of the level line so that it's approximately 4 in. above grade (the ground) at the highest corner. Place the point of a 2½-in.-long nail precisely on the level line, then push the nail—known as a grade nail—through the fiber-form tube (2). Repeat to install a grade nail in the five remaining tubes.

Shoot a level line across the tubes with a laser level. Press a nail through the tubes on the line to indicate the height of the concrete pier.

CEDAR-SHINGLE SHED

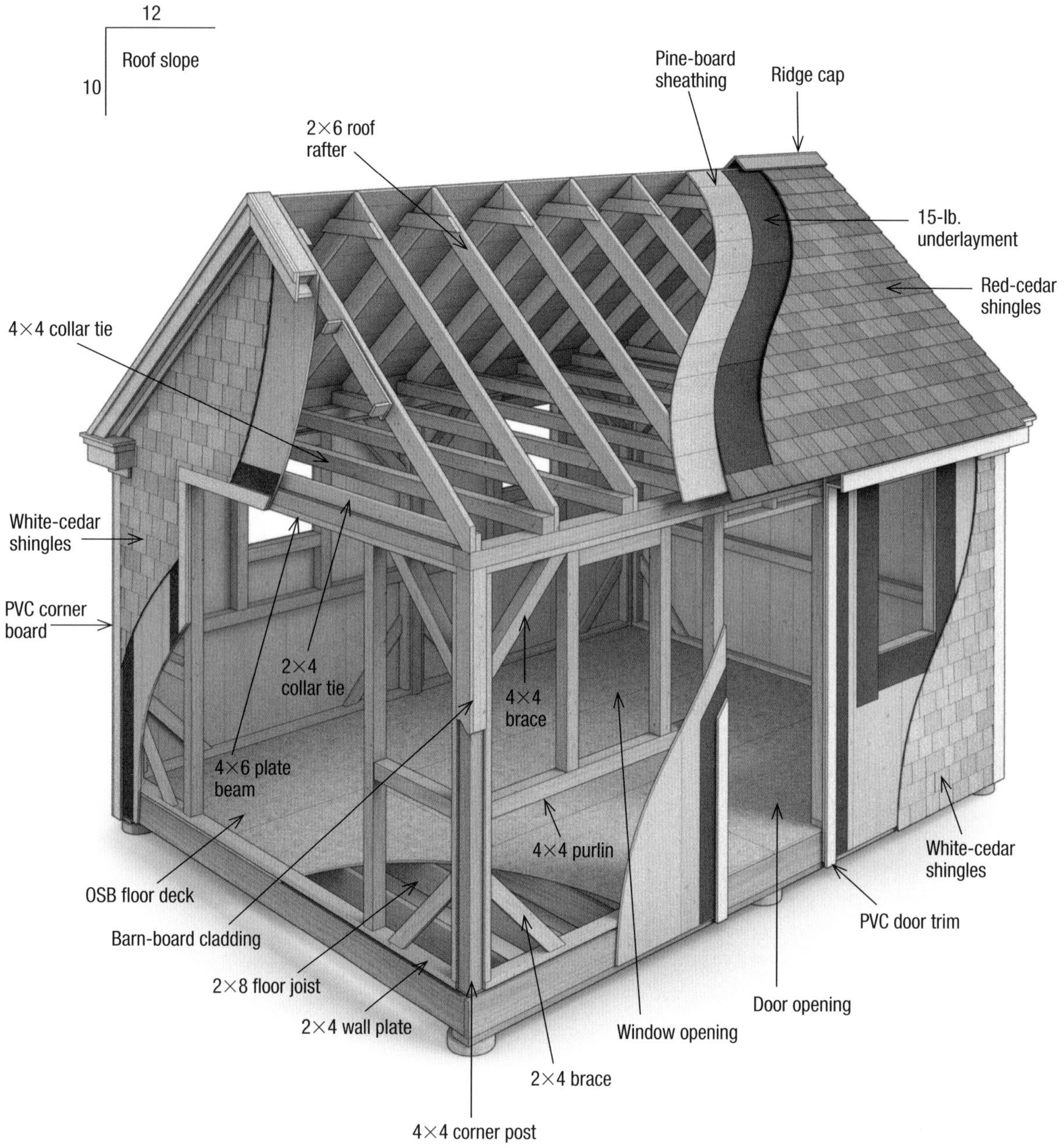

12

Roof slope

10

2×6 roof rafter

Pine-board sheathing

Ridge cap

15-lb. underlayment

Red-cedar shingles

4×4 collar tie

White-cedar shingles

PVC corner board

2×4 collar tie

4×6 plate beam

4×4 brace

4×4 purlin

White-cedar shingles

OSB floor deck

Barn-board cladding

2×8 floor joist

2×4 wall plate

2×4 brace

PVC door trim

Door opening

Window opening

4×4 corner post

Insert a 10-in. J-bolt into the wet concrete. Position the bolt 3½ in. from the outer edge of the fiber-form tube and leave 3 in. protruding.

To prevent weed growth and erosion, spread 2 in. to 4 in. of gravel across the site. Rake the gravel 12 in. to 18 in. beyond the concrete piers.

Now use a hoe or shovel to mix together water and dry concrete in a wheelbarrow. Or rent an electric cement mixer. It typically takes about 3 qt. water to fully hydrate 80 lb. of concrete. The amount of concrete needed depends on the depth of the holes. As a general rule, it takes about three 80-lb. bags of concrete to fill a 10-in.-dia. by 36-in.-deep hole. If the holes are 42 in. deep, you'll need 3¼ bags per hole.

Slowly pour the concrete into each fiber-form tube. Stop when the top of the concrete pier is flush with the grade nail. If you accidentally add too much concrete, use a small trowel to remove the excess concrete and expose the nail. Before moving on to the next step, double-check to ensure that each fiber-form tube is filled flush with its grade nail.

Next, insert a ⅝-in.-dia. by 10-in.-long J-bolt into the top of each concrete pier. Be sure the short right-angle leg at the end of each bolt is pointing toward the center of the pier (**3**). Leave approximately 3 in. of the bolts protruding and position each one 3½ in. from the outer edge of the fiber-form tubes. Positioning the bolts off-center keeps most of the concrete piers tucked under the floor frame.

After installing all six J-bolts, spread 2 in. to 4 in. of gravel across the building site to deter weed growth and prevent erosion (**4**). Allow the concrete to cure overnight, then peel off the fiber-form tubes to reveal the concrete piers (**5**).

After allowing the concrete to cure overnight, peel the fiber-form tubes from the piers and then cut flush with the gravel bed.

SAFETY FIRST

• Before digging the pier holes, have the local utility company come out and check the area for any buried cables, pipes, or gas lines.

• When mixing and pouring concrete always wear work gloves and eye goggles. Wet concrete is caustic and can burn your skin and eyes.

CONCRETE-PIER LAYOUT

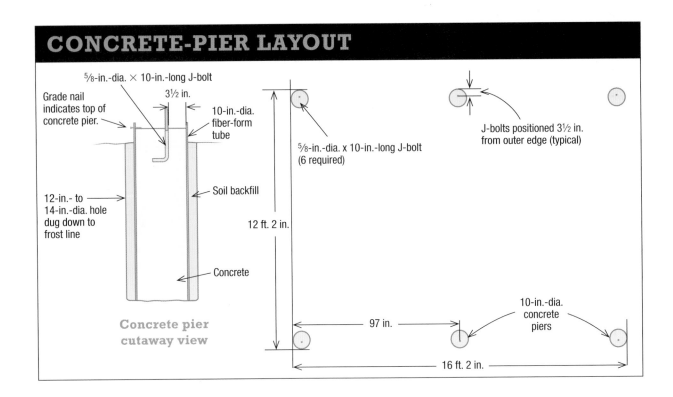

⅝-in.-dia. × 10-in.-long J-bolt

Grade nail indicates top of concrete pier.

3½ in.

10-in.-dia. fiber-form tube

⅝-in.-dia. x 10-in.-long J-bolt (6 required)

J-bolts positioned 3½ in. from outer edge (typical)

12-in.- to 14-in.-dia. hole dug down to frost line

Soil backfill

12 ft. 2 in.

Concrete

Concrete pier cutaway view

10-in.-dia. concrete piers

97 in.

16 ft. 2 in.

BUILD THE FLOOR FRAME

Start by cutting two pressure-treated 2×6 mudsills and two pressure-treated 2×8 rim joists to 16 ft. long each. Set one 2×6 mudsill atop each row of concrete piers. Confirm that the ends of the sills extend past the bolts in the corner piers by an equal amount. Now mark where the center of each J-bolt meets the mudsill (1). Place the mudsills on sawhorses and drill a 1-in.-dia. hole through the center of the 2×6 sill at each J-bolt location.

Temporarily set the mudsills back on top of the concrete piers with the J-bolts protruding through the 1-in. holes. Check the mudsills for

Set a 2×6 mudsill on top of the concrete piers. Mark the location of the J-bolts onto the sill, then drill a 1-in.-dia. hole at each mark.

Check the mudsills for square by measuring the opposing diagonals. When the two dimensions are equal, the sills are square.

3

Fasten the 2×6 mudsill to the 2×8 rim joist with 3½-in.-long nails. Space the nails 12 in. to 16 in. apart along the length of the 16-ft. sill.

4

Look along the edge of each pressure-treated 2×8 floor joist. If it has a slight crown in it, install the joist with the crown facing up.

square by pulling two tape measures diagonally across opposing corners (**2**). Shift the 2×6s until the two dimensions are equal, meaning that the sills are square. If necessary, enlarge one or more bolt holes to permit squaring up the two sills.

Remove the mudsills and nail each one to a 16-ft.-long 2×8 rim joist (**3**). Set the L-shaped sills back on top of the piers with the J-bolts poking through the holes. Next, cut pressure-treated 2×8 floor joists to fit between the rim joists. Space the joists 16 in. on center (**4**).

Check the floor frame for square one last time, then drive three 3½-in.-long nails through the rim joists and into each end of every 2×8 floor joist. Place a washer and hex nut onto the J-bolts and tighten the nuts with a wrench (**5**). Cover the floor frame with ⅝-in.-thick exterior-grade oriented-strand board (**6**). Leave a ⅛-in. gap between the sheets for expansion, then secure them to the joists with 2½-in.-long ring-shank nails spaced 10 in. to 12 in. apart.

HELPFUL HINTS

• Drill an oversized 1-in.-dia. hole for each ⅝-in.-dia. J-bolt and you'll create a little extra wiggle room for shifting the mudsills into position.

• Extend the gravel bed 12 in. to 18 in. beyond the concrete foundation piers. That way, it'll help drain away rainwater from the perimeter of the shed and eliminate soggy, muddy spots.

Square up the floor frame, then put a washer and nut onto each J-bolt. Tighten the nuts with a wrench to secure the frame to the concrete piers.

Cover the shed's floor frame with ⅝-in.-thick oriented-strand board (OSB). Fasten the OSB sheets with 2½-in.-long ring-shank nails.

5

6

Use the miter saw to simultaneously cut two 2×4 wall plates. Gang-cutting multiple parts saves time and reduces measuring mistakes.

Build the timber-panel wall frame out of 2×4s. Then, fasten a 4×4 post in between the top and bottom wall plates with 3-in. screws.

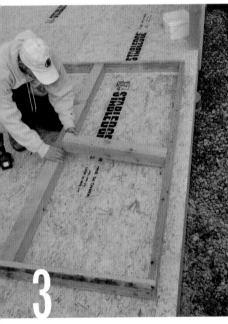

Cut a 4×4 purlin to fit between the 2×4 end stud and 4×4 post. The purlin strengthens the wall and provides nailing support for the pine sheathing.

TOOL TIP

Save time and increase accuracy by using the miter saw to gang-cut two or more boards to length at the same time.

Wall Framing

As mentioned earlier, each timber-panel wall consists of a 2×4 frame and 4×4 posts, purlins, and braces. The front sidewall has rough openings for a 3-ft.-wide door and two 25¼-in.-wide by 41-in.-tall double-hung windows. The rear sidewall has openings for two 24-in.-sq. awning windows. The left-side gable-end wall has a 6-ft.-wide opening for a pair of swinging doors. The opposite gable-end wall has no windows or doors, leaving plenty of interior wall space for mounting shelves and hanging tools.

BUILD THE GABLE-END WALL

Begin by cutting the parts to build the gable-end wall with the double-door opening. Use a power miter saw to simultaneously cut two 2×4s to 11 ft. 3 in. long (**1**). These boards will form the top and bottom wall plates. Then cut two 2×4 studs and two 4×4 posts to 80½ in. long. Set the studs between the top and bottom wall plates, positioning one at each end to form a rectangular frame. Use an impact driver to drive 3-in. screws through the plates and into the ends of the studs.

Now position one 4×4 post 26¼ in. away from each end stud. These two posts create the rough opening for the double doors. (Refer to the Wall Framing Detail drawing on the facing page.) Screw the posts between the wall plates (**2**). Then, cut two 4×4 purlins to 26¼ in. long and install each between an end stud and post (**3**).

Next, rotate the miter-saw blade to 45° and miter-cut two 2×4 diagonal braces to 30 in. Then miter-cut two 4×4 diagonal braces to 32 in. Set the 4×4 braces into the upper corners of the wall frame and

Fasten 4×4 diagonal braces to the upper corners of each timber-panel wall section. Secure each brace end with two 3-in. screws.

Cut and install 2×4 diagonal braces to the lower corners of each timber-panel wall. Set the braces flush with the wall's outer surface.

Nail ⁷⁄₈-in.-thick pine barn boards to two sides of each 4×4 corner post. Fasten the pine boards with 2¹⁄₂-in.-long (8d) nails.

secure each end with two 3-in. screws **(4)**. Fasten the 2×4 braces to the bottom corners of the wall frame. Again, secure each end with two 3-in. screws **(5)**.

Now make four corner posts, one for each corner of the shed floor. Start by cutting four 4×4s to 83¹⁄₂ in. long. Then, from ⁷⁄₈-in.-thick pine barn board cut four 3¹⁄₂-in.-wide by 83¹⁄₂-in.-long boards, and four 4³⁄₈-in.-wide by 83¹⁄₂-in.-long boards. Nail one 3¹⁄₂-in.-wide barn board to each 4×4 post, then nail on the second barn board, overlapping the edge of the first board **(6)**.

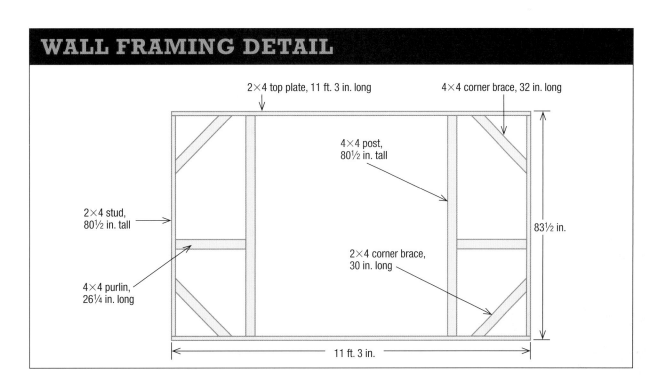

WALL FRAMING DETAIL

2×4 top plate, 11 ft. 3 in. long

4×4 corner brace, 32 in. long

4×4 post, 80¹⁄₂ in. tall

2×4 stud, 80¹⁄₂ in. tall

2×4 corner brace, 30 in. long

4×4 purlin, 26¹⁄₄ in. long

83¹⁄₂ in.

11 ft. 3 in.

1 Stand the corner post with its two padded sides flush with the outer edges of the shed's floor. Then toe-nail the post to the floor deck.

2 Fasten the timber-panel wall frame to the corner post with 2½-in.-long washer-head screws. Space the screws 14 in. to 16 in. apart.

3 Stand another 4×4 corner post at the opposite end of the wall frame. Secure the post to the 2×4 end stud with 2½-in. screws.

HELPFUL HINT

The pine-board sheathing can be nailed to the wall frames while the frames are lying flat on the floor deck, or after they've been tipped up into place. Either method is acceptable, but if you're hand-nailing the sheathing, as opposed to using a pneumatic nailer, then on-deck nailing is easier and faster.

The purpose of the barn boards is to pad out the 4×4 post so that its interior corner will protrude past the 2×4 wall studs, creating a corner-post reveal. Without the ⅞-in.-thick padding, the 4×4 posts would be hidden behind the studs.

ERECT THE GABLE-END WALL

Stand one post on the very corner of the shed floor with the two padded sides facing out. Toe-nail the bottom of the post to the floor deck (1). Tilt up the wall frame and slide it against the corner post. Hold the outside of the post flush with the outside of the wall frame. Fasten the wall to the post with four evenly spaced 2½-in.-long washer-head screws (2). Use the same screws to secure the bottom 2×4 wall plate to the floor deck. Install another corner post against the opposite end of the wall frame (3). Again, keep the post flush with the outside of the wall and secure it with four 2½-in. screws.

CONSTRUCT THE REMAINING WALLS

Build and erect the remaining three walls, using the same timber-panel construction method, as described above. The opposite gable-end wall is identical to the first, except that it doesn't have any 4×4 posts. Instead, there's an 11-ft.-long horizontal 4×4 purlin that runs the length of the wall in between the end studs.

Build two identical timber-panel frames to form the 16-ft.-long rear sidewall. Make each frame from two 89⅞-in.-long wall plates, two 80½-in.-tall studs, and one 86⅞-in.-long 4×4 horizontal purlin. Cut a 4×4 center post to 83½ in. long and nail a ⅞-in.-thick by 3½-in.-wide barn board to one side of the post. Stand the post between the two timber-panel frames with its padded side facing out. Screw each wall

1 Set a center post between the two timber-panel frames that form the rear sidewall. Screw the wall frames to each side of the post.

2 Install a 4×6 plate beam on top of the shed walls. Position the outside of the beam flush with the outer edge of the 2×4 top wall plate.

3 Fasten together the 4×6 plate beams at the corners with 10-in. screws. Then, drive 2½-in. screws through the top wall plate and into the beam.

frame to the center post (**1**). Then install a double-padded corner post at each end of the rear sidewall.

The front sidewall is also composed of two timber-panel frames. However, unlike the rear sidewall the two frames aren't separated by a center post. Instead the two 73¼-in.-long wall frames are joined together by a 36½-in.-wide header that creates the rough opening for the single-width doorway.

Build each front sidewall timber-panel frame from two 73¼-in.-long wall plates, an 80½-in.-tall 2×4 stud, an 80½-in.-tall 4×4 post, and a 70¼-in.-long 4×4 horizontal purlin.

Once you've erected all four walls, cut a 4×6 plate beam to run along the tops of the walls. Cut two beams to 16 ft. long for each sidewall, and two beams to 11 ft. 1 in. for each gable-end wall. Set the 4×6 beams on the walls, flush with the outside edge of the 2×4 top plate (**2**). Fasten the beams together at each corner with two 10-in.-long structural screws (**3**). Then drive 2½-in.-long screws up through the 2×4 top wall plate and into the underside of the plate beam, spacing the screws 16 in. to 20 in. apart.

4 Set the door header between the two timber-panel sections in the front sidewall, then fasten the header to the plate beam with 2½-in. screws.

Now make the door header by first cutting two 2×4s to 36½ in. and three 2×4s to 2½ in. Nail the short 2×4 blocks between the two longer 2×4s, creating a 5½-in.-tall header. Set the header between the timber-panel frames in the front sidewall, and fasten it to the plate beam with four 2½-in. screws (**4**).

Next, sheathe the exterior of the shed walls with 1×10 and 1×12 pine boards. Nail the boards vertically to the top and bottom wall plates, horizontal purlins, and 4×4 posts. Use a circular saw to cut the sheathing from the rough window openings (**5**).

5 Use a portable circular saw to cut the rough window openings out of the pine sheathing. Adjust the saw for a ⅞-in.-deep cut.

Miter-cut the upper ends of each 2×6 roof rafter to 40° in order to create the proper 10-in-12 roof slope. Cut a total of 18 rafters.

Join together two rafters to create a roof truss. Then nail a 24-in.-long 2×4 gusset plate across the joint to lock the rafters together.

Note that you could save some time and money by sheathing the walls with ½-in.-thick exterior-grade plywood. But the pine boards are much more attractive and lend a more traditional look to the shed's interior.

Roof Framing

The shed's roof frame consists of 2×6 rafters and 4×4 collar ties. The upper ends of the rafters are cut to 40°, creating a 10-in-12 roof slope. The lower ends of each rafter are trimmed flush with the edge of the top wall plate.

To simplify the roof framing, we joined together two rafters with a 2×4 gusset plate to create a roof truss. However, unlike a typical roof truss, we didn't install a horizontal chord across the bottom of the truss. Instead, 4×4 collar ties are installed alongside each truss, spanning the sidewalls.

Once the nine trusses and collar ties are installed, the roof is sheathed with 1×10 and 1×12 pine boards. Again, you could sheathe the roof with plywood, but the pine boards are much more attractive from the inside.

BUILD THE TRUSSES

Begin by using the miter saw to trim the upper end of each 2×6 rafter to 40° (**1**). Measure 93¾ in. from the upper end and miter-cut the lower end of the rafter to 40°. Then, cut a 2-in.-long seat cut into the bottom of each rafter. (Refer to the Roof Truss drawing on the facing page.)

3 **Stand each roof truss** into place on top of the sidewalls. Space the trusses 24 in. apart and toe-nail each rafter end to the top plate.

Now cut nine 24-in.-long 2×4 gusset plates, mitering each end to 40°. Butt the upper ends of two rafters tightly together, and then nail on the gusset plate (2). Repeat to assemble the remaining eight trusses.

Set the trusses on top of the sidewalls and toe-nail the lower end of each rafter to the top wall plate (3). Space the trusses 24 in. on-center. Next, cut seven 4×4 collar ties and two 2×4 collar ties, making each 12 ft. long. Lay the 2×4 collar ties across the sidewalls at each end of the shed. Nail the 2×4 collar ties to the rafters of the gable-end trusses.

ROOF TRUSS

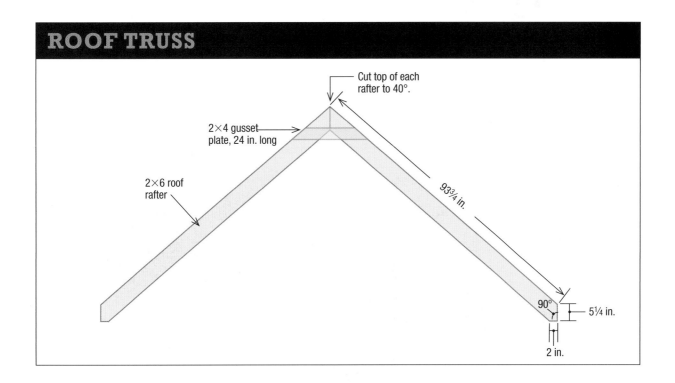

Cut top of each rafter to 40°.

2×4 gusset plate, 24 in. long

2×6 roof rafter

93¾ in.

90°

5¼ in.

2 in.

Fasten a 4×4 collar tie to the plate beam with 8-in. structural screws. Then, drive 3-in. screws through the rafters and into the collar tie.

Sheathe both gable-end trusses with 1×10 and 1×12 pine boards. Fasten the vertical-board sheathing with 2½-in. ring-shank nails.

Lay the 4×4 collar ties across the sidewalls, placing one against each of the remaining seven trusses. Fasten the end of each collar tie to the 4×6 plate beam with an 8-in.-long structural screw **(4)**. Then drive two 3-in. screws through each rafter and into the collar tie. After installing all the collar ties, sheathe the outer surface of each gable-end wall with 1×10 and 1×12 pine boards **(5)**. Fasten the sheathing with 2½-in.-long ring-shank nails. Then cover both sides of the roof with the same pine boards **(6)**.

INSTALL THE CORNICE

Since the roof rafters don't extend beyond the sidewalls, it's necessary to build a cornice, which forms an overhanging eave along each edge of the roof. The cornice consists of the pine frame covered with PVC trim boards. The pine frame is completely hidden from view once the cornice is installed, and the PVC trim creates a no-maintenance exterior. (See the Cornice Construction drawing on the facing page.) And note that the cornice extends beyond the sidewalls and returns back along each gable-end wall. These short cornice returns are stylish architectural features not normally found on a storage shed.

Start by making the 16-ft.-long backboard, which runs the length of the cornice, from a pine 1×8. Since you're not likely to find a 1×8 that long, you'll have to fabricate it from two boards. Tilt the tablesaw blade to 40° and bevel-rip the edge off the 1×8s. Then crank the blade back up to 0° and rip the backboard to 6¼ in. wide.

Now cut eight 8¾-in.-tall standoff blocks from a pine 1×6. Rip the 1×6 to 5 in. wide and then cut each 8¾-in.-tall block as shown in the drawing on the facing page: Trim the upper end to 40° and square-cut the lower end. Glue and nail the standoff blocks to the backboard, spacing them 24 in. apart.

Cover the roof with the same 1×10 and 1×12 pine boards used to sheathe the shed walls. Butt the boards together and nail them to the rafters.

CORNICE CONSTRUCTION

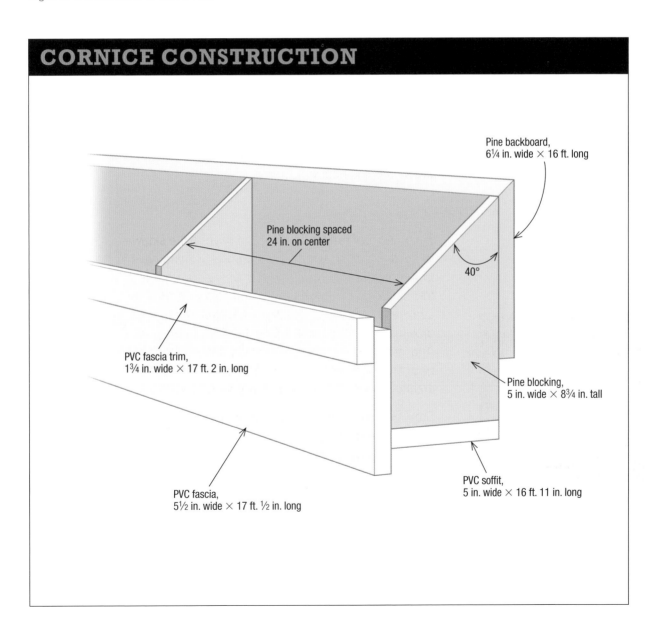

Pine backboard,
6¼ in. wide × 16 ft. long

Pine blocking spaced
24 in. on center

40°

PVC fascia trim,
1¾ in. wide × 17 ft. 2 in. long

Pine blocking,
5 in. wide × 8¾ in. tall

PVC fascia,
5½ in. wide × 17 ft. ½ in. long

PVC soffit,
5 in. wide × 16 ft. 11 in. long

Frame the cornice out of pine, then add the PVC-board exterior. Fasten the PVC fascia trim to the PVC fascia with 1½-in.-long nails.

Hold the cornice flush with the top of the rafters. Drive three 3-in. nails through the pine backboard and into the end of each rafter.

Next, nail a 5-in.-wide by 16-ft. 11-in.-long PVC soffit to the underside of the cornice. Then fasten a 5½-in.-wide by 17-ft. ½-in.-long PVC fascia across the front of the cornice. Finally, nail a 1¾-in.-wide by 17-ft. 2-in.-long PVC fascia trim to the fascia board (1).

Be sure the fascia trim extends ¾ in. above the top edge of the fascia board. Note that PVC trim boards are available in 18-ft. lengths, so you can cut each of these cornice pieces from a single board.

Hold the cornice in place with the backboard's upper edge flush with the top edge of the roof rafters. Secure the cornice by driving three 3-in. nails through the backboard and into each rafter (2). Cover the cornice by installing the final pine boards along the lower edge of the roof. Build and install a cornice onto the opposite side of the shed.

Now use the same cornice-building technique to fabricate four 16-in.-long cornice returns. Fit the returns tightly against each end of both cornices, then nail them to the shed wall. Cut a 7-in.-wide by 18-in.-long PVC cap and nail it on top of each cornice return (3).

Weatherproof the cornice by covering the PVC cap with aluminum flashing. Cut a 10-in. by 19-in. piece of flashing and bend it lengthwise to create a 4-in. by 6-in. L-shaped piece. Set the flashing on top of the cornice return with the 6-in. leg down flat against the cap. Snip and bend the flashing to wrap around the edge of the roof. Fasten the flashing by stapling through the 4-in.-tall vertical leg (4). Don't nail or staple the flashing to the PVC cap.

HELPFUL HINT

PVC trim is made from polyvinyl chloride, a resilient plastic, making it ideal for use outdoors: It'll never rot, crack, or need painting. However, if you'd prefer to trim out the shed with real wood, consider red cedar or redwood. Both species are naturally resistant to decay and wood-boring bugs.

3 Nail a PVC cap on top of each cornice return. Cut the cap to extend ¾ in. beyond the narrow fascia trim nailed to the fascia board.

4 Staple a piece of aluminum flashing over the cornice-return cap. Be sure to staple only into the vertical flange, not into the horizontal cap.

BUILD THE GABLE OVERHANG

The roof extends past each end of the shed, a detail known as a gable overhang. However, rather than building out the gable-end trusses with fly rafters, we took a slightly different, and easier, approach. We constructed a simple pine frame, not unlike the cornice frame, and nailed it to each gable end. The frames, called flying rigs, are then wrapped in PVC trim boards. You'll need to build two flying rigs for

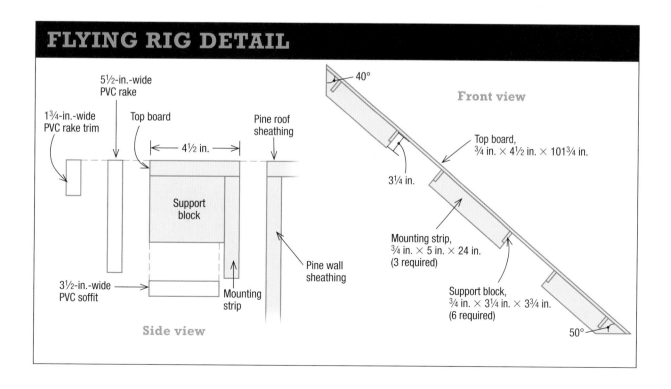

FLYING RIG DETAIL

1¾-in.-wide PVC rake trim

5½-in.-wide PVC rake

Top board

Pine roof sheathing

4½ in.

Support block

Pine wall sheathing

3½-in.-wide PVC soffit

Mounting strip

Side view

40°

Front view

3¼ in.

Top board, ¾ in. × 4½ in. × 101¾ in.

Mounting strip, ¾ in. × 5 in. × 24 in. (3 required)

Support block, ¾ in. × 3¼ in. × 3¾ in. (6 required)

50°

Hold the top of the flying rig flush with the roof sheathing, then nail through the mounting strips to secure the rig to the gable-end wall.

Fasten a 3½-in.-wide PVC trim board to the underside of the flying rig. Nail the ¾-in.-thick trim board to each pine support block.

Cut each rake board from a length of 1×6 PVC trim. Miter the upper end to 40° and the lower end to 50°. Nail the rake to the flying rig.

Fasten the 1¾-in.-wide PVC fascia trim to the fascia board. Hold the top of the narrow trim piece flush with the pine roof sheathing.

each gable-end wall. Each rig consists of 10 pine parts: a continuous top board, three mounting strips, and six short support blocks. (See the Flying Rig Detail drawing on p. 159.)

Start by using a tablesaw to rip a 10-ft.-long pine 1×6 to 4½ in. wide. Next, use a miter saw to bevel the upper end of the board to 40°. Measure 101¾ in. from the upper end and bevel-cut the board's lower end to 50°. This board will form the top of the flying rig.

Next, rip a pine 1x6 to 5 in. wide. Then use the miter saw to cut three 24-in.-long mounting strips. Miter-cut one end of the upper mounting strip to 40° and one end of the bottom strip to 50°. Square-cut both ends of the middle mounting strip.

Glue and screw the 4½-in.-wide top board to the mounting strips, making sure the 40° cut is flush at the upper end and the 50° cut is flush at the bottom end. Fasten the middle mounting strip in the center of the two end strips.

Now cut six ¾-in.-thick pine support blocks, making each one 3¼ in. wide by 3¾ in. long. Glue and screw the blocks to the flying rig, as shown in the drawing on p. 159.

To install the flying rig, hold it against the gable-end wall with its top flush with the roof sheathing **(1)**. Fasten the rig by driving six nails through each mounting strip. Repeat to build and install the remaining flying rigs.

Next, cut a 3½-in.-wide PVC soffit to fit against the underside of the flying rig. Bevel the board's upper end to 40° and its bottom end to 50°. Nail the board to each support block **(2)**. Miter-cut a 5½-in.-wide PVC rake board and fasten it to the front of the flying rig **(3)**. Then nail a 1¾-in.-wide strip of PVC rake trim to the rake board **(4)**. Repeat the previous steps to trim out the remaining flying rigs.

Staple 9-in.-wide strips of asphalt-saturated underlayment around the window openings. Be sure the side strips overlap the bottom strip.

Hold a 4-ft. or 6-ft. level against the window to make sure it's perfectly plumb, then fasten the window to the shed wall with 3½-in. nails.

INSTALL THE WINDOWS AND FINAL TRIM

Use a utility knife to cut 9-in.-wide strips of 15-lb. asphalt-saturated underlayment, also called builder's paper. Staple one strip along the bottom of each window opening, then fasten one strip to each side of the openings (1). Be sure the vertical side strips overlap the horizontal bottom strips. Also, staple 9-in. underlayment strips around the doorway openings and wrap every shed corner with 12-in.-wide strips.

Set the window into the rough opening, check it for level and plumb, and then nail it to the shed wall (2). Use the same procedure to install the remaining windows.

Next, make corner boards by cutting four 1×4 PVC boards and four 1×6 PVC boards to 91 in. long each. Nail the 1×4s to the gable-end walls, holding them perfectly flush with the shed corner. Then install the 1×6s against the sidewalls, making sure they overlap the 1×4s (3). Also, trim out the doorways with 1×4 PVC boards.

Cut the corner boards from PVC trim. Nail the first corner board to the gable-end wall. Install the second corner board to the sidewall.

HELPFUL HINT

Close and lock the window sash before setting the window into the rough opening. That'll help keep the window frame square.

Fasten a 1×2 ledger strip along the bottom of the wall, then install two rows of shingles, one on top of the other, to create the starter course.

Raise the ledger 5¼ in. to establish the shingles' proper exposure to the weather. Check to be sure the ledger is perfectly level.

Cedar-Shingle Siding

Following traditional practices, we installed eastern white-cedar shingles on the walls and western red-cedar shingles on the roof. White cedar is slightly thinner and less weather-resistant than red cedar, so you wouldn't use it on a roof. But white cedar is perfectly suitable for vertical applications, and it costs less than red cedar.

SHINGLE THE WALLS

Begin by tack-nailing a 1×2 ledger board to the bottom of the wall. Position the top of the ledger flush with the bottom of the PVC trim boards at the shed corner and doorway. Check the ledger for level, then install the starter course, which is two layers of white-cedar shingles. The first layer is called the under course.

Stand two or three shingles on the ledger and fasten them to the wall. Here, we used a ⁷/₁₆-in.-crown pneumatic stapler and 1-in. staples, but you could hand-nail them on with shingle nails, as well. Place the staples or nails ¾ in. from each edge and 6¼ in. up from the butt (bottom) end. That way, the shingle course above will overlap the fasteners by 1 in.

Use two fasteners per shingle, regardless of its width, and leave a ⅛-in. space—called a keyway—between each shingle. When you must trim a shingle to fit, score it with a sharp utility knife and snap it in two. Then nail another row of shingles right on top of the under course (**1**). Overlap all vertical seams between the shingles by at least 1½ in.

Stand the shingles on the ledger, spaced ⅛ in. apart. Stagger all seams by at least 1½ in., then secure each shingle with two staples.

Cut and install a short 1×2 ledger strip to fit beside the window. Measure carefully to maintain the 5¼-in. exposure to the weather.

After completing the double-row starter course, move the ledger up 5¼ in. to create the proper exposure to the weather **(2)**. Staple on the second shingle course. Again, be sure to leave a ⅛-in. gap between shingles, stagger all vertical seams by at least 1½ in., and secure each shingle with two fasteners placed ¾ in. from each edge **(3)**. Continue shingling up the walls of the shed, one course at a time. When you reach the windows, cut a short ledger to fit between the window and the corner boards **(4)**.

Cedar Roofing

As mentioned earlier, the roof is covered with western red-cedar shingles, which are more weather resistant than eastern white cedar. The installation of the roof shingles is similar to shingling the walls, with three exceptions: The roof deck is protected with a layer of 15-lb. underlayment, the shingles' exposure to the weather is 6 in., and the shingles are secured with 1½-in.-long stainless steel staples.

INSTALL A STARTER COURSE

Begin by installing a double-row starter course of shingles along the lower edge of the roof. Allow the butt end of the under-course shingles to extend beyond the roof edge so they're 1½ in. from the PVC fascia trim. Leave a gap of ¼ in. to ⅜ in. between the shingles. Fasten each shingle with two staples positioned ¾ in. from each edge and 7 in. up from the butt end **(1)**. Install another row of shingles right on

Install an under course of red-cedar shingles along the roof's edge. Secure each shingle with two staples placed ¾ in. from each edge.

Double-up the under course by installing a second row of shingles right on top. Leave a gap of ¼ in. to ⅜ in. between the shingles.

Cover the roof with 15-lb. underlayment. Position the 3-ft.-wide sheet 6¾ in. above the butt edge of the shingles in the starter course.

SAFETY FIRST

For maximum safety, be sure to nail roofing brackets into a rafter, not just into the roof sheathing.

top of the under course. Keep the butt ends flush and maintain the proper expansion space between each shingle **(2)**. Stagger all vertical seams between the two rows by at least 1½ in.

Unroll 15-lb. underlayment across the roof **(3)**. Place it 6¾ in. up from the butt end of the starter course so it'll conceal the staples but still be overlapped by the next shingle course. Secure the underlayment with ³⁄₁₆-in. staples.

CONTINUE ROOFING

Tack-nail a 1×3 ledger board to the roof, positioning its upper edge exactly 6 in. from the butt end of the starter course. Lay out several shingles across the ledger, checking to make sure each vertical seam overlaps the seams in the course below by at least 1½ in. Space the shingles ¼ in. to ⅜ in. apart, then secure each one with two staples **(1)**.

Tack-nail a 1×3 ledger to the roof. Arrange several shingles along the ledger, spaced ¼ in. to ⅜ in. apart. Offset the seams by at least 1½ in., then staple down the shingles.

Snap a chalkline across the roof precisely 6 in. above the butt end of the second-row shingles. Move the ledger up to the chalkline.

Continue to install red-cedar shingles along the ledger. Maintain the 6-in. exposure and fasten every shingle with two 1½-in. staples.

Measure 6 in. up from the butt end of the second course and snap a chalkline across the roof **(2)**. Move the ledger up and tack-nail it flush with the chalkline. Stand several shingles on top of the ledger and continue stapling them down **(3)**.

After completing five or six courses, install roof brackets and staging planks so you can safely reach the upper half of the roof. Cover the exposed roof deck with 15-lb. underlayment, making sure to overlap the staples in the course below **(4)**. Continue shingling up the roof. Allow the shingles to extend past the peak, then trim them flush with a circular saw **(5)**. Repeat the previous steps to shingle the opposite roof plane.

Unroll 15-lb. underlayment across the exposed roof deck. Be sure to overlap the staples in the previous course of shingles.

Complete the last course by installing full-size shingles along the roof peak. Then use a circular saw to trim the shingles flush.

Staple a strip of 15-lb. underlayment to the underside of the red-cedar ridge cap. The asphalt-saturated strip will block out moisture.

Screw the ridge cap to the roof peak, then use a circular saw to trim the cap so that it extends 1 in. past the gable end of the roof.

INSTALL THE RIDGE CAP

Make a red-cedar ridge cap to cover the shingles along the roof peak. You'll need four cedar 1×6s: two 8-ft.-long and two 10-ft.-long boards.

Tilt the tablesaw blade to 10°, then bevel-rip an edge off one 8-ft. and one 10-ft. board. Remove only enough wood to bevel the edges, so the boards remain about 5½ in. wide. Now adjust the saw fence 4¾ in. from the blade and bevel-rip the remaining 8-ft. and 10-ft. cedar boards. Use 1⅝-in.-long stainless steel trim-head screws to fasten a 5½-in.-wide by 8-ft. board to a 4¾-in.-wide by 10-ft. board. Be sure to butt together the beveled edges. Then, screw on the remaining two boards, overlapping the end joints by 2 ft.

Next, slice a continuous piece of 15-lb. underlayment and staple it to the underside of the ridge cap **(1)**. Set the cap on top of the roof peak and secure it with 2½-in. stainless-steel trim-head screws, spaced 16 in. to 20 in. apart. Then use a circular saw to trim the ends of the ridge cap so they extend 1 in. past the roof shingles **(2)**.

Frame-and-Panel Doors

The shed has a single door on the front sidewall and a pair of doors on the gable end. All three doors are the same size: 36 in. wide by 79 in. tall. And each has a 5/4 pine frame and ¾-in.-thick pine panel, which is composed of 1×6 tongue-and-groove beadboard planks. Here, we'll show how to build the single door, but the other two are identical, except for the paint color.

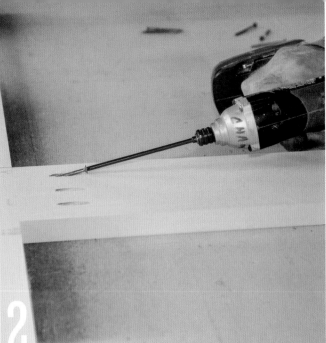

Assemble the pine door frame by setting the three 25-in.-long horizontal rails in between the two 79-in.-tall vertical stiles.

Fasten the three door rails to the stiles with 2½-in.-long pocket screws. Drive three screws through each end of every horizontal rail.

MAKE THE DOOR FRAME

The door's frame consists of five parts: two 79-in.-tall vertical stiles and three 25-in.-long horizontal rails. Crosscut the parts to length from 5/4 by 5½-in.-wide pine boards.

Next, use a pocket-hole jig to drill three pocket-screw holes into both ends of each rail. Lay out the frame on a workbench with the rails between the stiles **(1)**. Position the upper and lower rails flush with the ends of the stiles. Place the middle rail 40 in. from the bottom of the door. Fasten the rails to the stiles with 2½-in.-long pocket screws **(2)**.

MAKE THE DOOR PANEL

Now construct the door panel by cutting seven 1×6 tongue-and-groove pine beadboard planks to 79 in. long each. Use the tablesaw to rip one of the planks into two narrow strips, which will be installed at the right and left edges of the door.

Lock the tablesaw fence 2 in. from the blade. Place the groove edge of one beadboard plank against the fence and rip it to width. Adjust the saw fence for a 2¾-in. cut. Now set the tongue edge of a plank against the fence and rip it to width. Before assembling the door panel, brush a coat of paint onto the tongue edge of every plank **(1)** (see p. 168).

Set the 2¾-in.-wide beadboard strip flush along the edge of the door's frame with its tongue edge facing toward the center of the frame. Fasten the strip by driving 1½-in.-long staples through the tongue and into the pine frame. Now install six full-width beadboard planks, stapling each one through the tongue edge **(2)** (see p. 168). Fasten the 2-in.-wide beadboard strip last.

> **HELPFUL HINT**
>
> Cut a 15° bevel into the top edge of the bottom and middle door rails to help shed rain and snow.

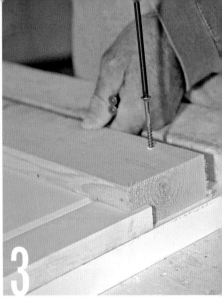

Paint the tongue edge of the bead-board planks before assembling the door. That way, if the planks shrink, you won't see vertical stripes of bare wood.

Staple the tongue-and-groove planks to the door frame. Drive the 1½-in.-long staples at an angle through the exposed tongue edge.

Screw a 2×4 stiffener to the top and bottom of the door. Note how the ¾-in.-deep rabbet cut in the 2×4 overlaps the beadboard panel.

Next, cut two 35½-in.-long door stiffeners from a 2×4. Use the tablesaw to cut a ¾-in.-deep rabbet into each stiffener. Lay the stiffeners across the top and bottom of the door with the rabbets overlapping the door panel. Fasten the stiffeners with 2-in.-long screws **(3)**. Apply two coats of exterior-grade paint to both sides and all edges of the door.

INSTALL THE DOOR

Screw three 10-in. or longer strap hinges to the door, then tip the door into place **(1)**. Use shims to center the door within the opening. Secure each hinge leaf by screwing through the PVC door casing and wall sheathing and into the 4×4 post **(2)**.

Mount three strap hinges to the door, then tip the door into the opening. Insert shims around the door to create an equal gap on all sides.

Secure the door hinges to the shed by driving 3-in. screws through the PVC trim and pine sheathing and into the 4×4 wall post.

Build a platform step with pressure-treated 2×8s. Make the frame 16 in. wide by 40 in. long, with a 2×8 center support in the middle.

Top the frame with three 41-in.-long pieces of pressure-treated decking. Secure the decking with 2½-in.-long screws or nails.

BUILD STEPS AND RAMPS

Construct a small wooden platform for placement in front of the single door. Cut and build a 16-in.-wide by 40-in.-long frame out of pressure-treated 2×8s. Install a 2×8 center support across the middle of the frame (**1**). Then cover the frame with 5/4 by 6-in. pressure-treated decking (**2**). Set the platform in front of the door (**3**). Check it for level, then drive screws at an angle through each end of the platform and into the shed wall.

We also built a pair of ramps for the double doors on the gable end. Make each ramp 35 in. wide and 48 in. to 54 in. long. Cut 1×10 and 1×12 pine boards for the top surface of the ramps, then screw the boards to 35-in.-long 2×4 cleats. Use two cleats for each ramp.

Screw a 1×4 ledger to the shed directly below the double-door threshold. Position the top of the ledger 1 in. below the shed floor. Install the ramps by simply setting them onto the ledger (**4**). You can screw the ramps to the ledger if you'd like, but leaving them unattached makes it easy to move them out of the way when necessary.

Set the platform step on the gravel bed in front of the shed door. Check it for level, then toe-screw the platform to the shed wall.

Construct a pair of removable ramps for access through the double doors. Rest the ramps on a 1×4 ledger screwed below the threshold.

DESIGN DETAILS

1

2

1. The rear sidewall has two 24-in.-sq. awning windows that admit natural light and fresh air. The roof is covered with red-cedar shingles.

2. The beauty of timber-panel construction is evident on the interior, with exposed posts and beams and wide-board roof and wall sheathing.

3. The 3-ft.-wide frame-and-panel door is painted a coastal-blue color, in keeping with tradition. It's flanked by double-hung windows.

4. The gable end of the 12-ft.-wide shed features white-cedar shingles and a doublewide doorway with wooden ramp for easy, roll-in access.

5. Gable overhangs and cornice returns add visual interest and a little classical architecture to this quaint cedar-shingled storage shed.

3

4 5

6

POST-AND-
BEAM BARN

This spacious 14-ft. by 20-ft. post-and-beam barn melds traditional barn architecture with modern building methods. And the result is a large timber-frame building that goes up surprisingly fast. The barn features a white-pine frame of rough-sawn 6×6 posts, 6×10 beams, and 4×8 rafters. The walls are braced with diagonal 4×5s, and the window openings are framed with 4×4s.

However, unlike a traditional post-and-beam structure, there's not a single mortise-and-tenon joint or scarf joint in the whole barn. Instead, we assembled the frame with metal fasteners called T-Rex connectors. The T-shaped fasteners were screwed in place and then slid into slots cut into the ends of the posts and beams. Aluminum pins were driven through holes bored in the timbers and fasteners to securely hold together the barn frame. Now this modern construction method might not impress timber-frame purists, but it does provide a quick and easy way for do-it-yourselfers to build a beautiful post-and-beam barn.

Other notable features include vertical-board pine siding, two round windows, an interior storage loft with ladder, architectural-style asphalt roof shingles, and two styles of doors: an outswinging hinged pair on the front gable end and a traditional rolling barn door on the sidewall.

And while it's attractive on the outside, what makes this barn truly special is what's visible on the inside: an exposed frame of large white-pine timbers that would warm the heart of any barn builder. (To order building plans for the Post-and-Beam Barn, see Resources on p. 214.)

Skid-Frame Foundation

The local building department allowed us to build this 280-sq.-ft. barn on an on-grade foundation, meaning that we didn't have to dig down to the frost line. Permanent, frost-proof foundations are typically required only for structures over 400 sq. ft., but that requirement varies from town to town, so be sure to check the building code in your area before proceeding.

Here, we built a skid foundation made up of long wood beams (skids) laid across rows of solid-concrete blocks. Skids are traditionally made of solid-wood timbers, such as 4×6s, 6×6s, or even 8×8s, but solid timbers tend to bow, crack, and warp over time. Plus, it's hard to find really long timbers that are perfectly straight. We eliminated those problems by fabricating each of the three 20-ft.-long skids out of pressure-treated 2×6s.

SET THE SKIDS

Start by laying out three parallel rows of 2-in.-thick solid-concrete blocks. Place five equally spaced blocks in each row, and use a long 2×6 board to align the blocks in straight rows (1). Make each five-block row 19 ft. 6 in. long. Space the first and third rows precisely 12 ft. 11 ½ in. apart, as measured from the center of one row to the center of the other. Center the middle row of blocks in between the other two rows.

To see if a row of blocks is level, set the long board on edge on the blocks. Place a 4-ft. level on top of the long board and check for level. If necessary, shim up the lowest block by adding another solid-concrete block, short lengths of pressure-treated lumber, or strips of asphalt roofing. Our building site was relatively level, so we didn't have to do much shimming. However, if your barn site slopes down dramatically, build up the low end with 4-in.-thick or even 8-in.-thick solid-concrete blocks.

Once the foundation blocks are in position, make the three skids from 20-ft.-long pressure-treated 2×6s. Form each skid by fastening together three 2×6s with 3-in.-long decking screws (2). Drive in a row of the screws from each side of the skid in a zigzag pattern; space the screws about 12 in. apart.

Set the solid-concrete blocks into place, then check them for level. Use a long, straight 2×6 to align each row of five concrete blocks.

Make each 20-ft.-long skid by screwing together three pressure-treated 2×6s. Drive the 3-in. screws in from both sides of the skid.

TIMBER FRAME

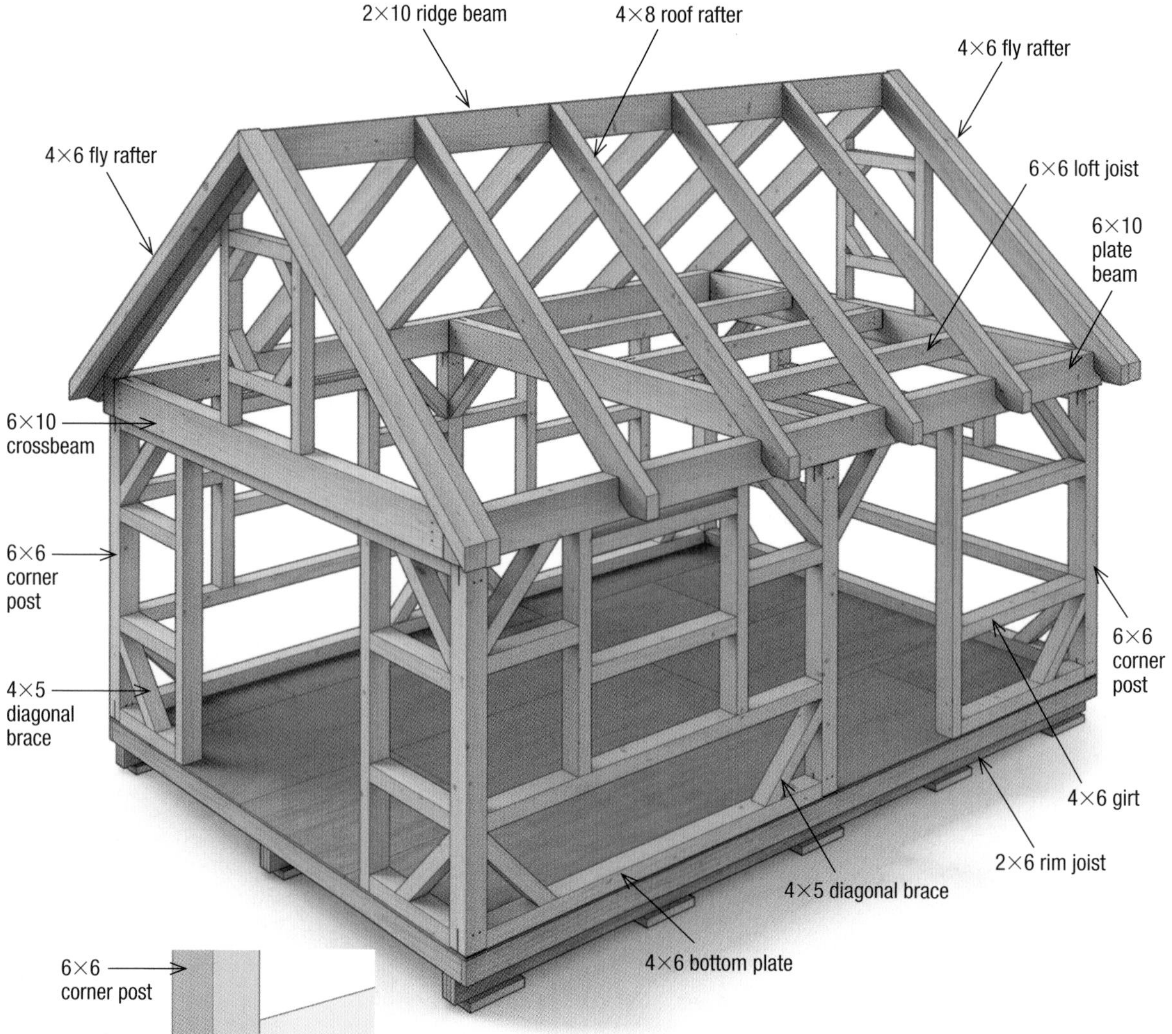
2×10 ridge beam
4×8 roof rafter
4×6 fly rafter
4×6 fly rafter
6×6 loft joist
6×10 plate beam
6×10 crossbeam
6×6 corner post
6×6 corner post
4×5 diagonal brace
4×6 girt
2×6 rim joist
4×5 diagonal brace
4×6 bottom plate

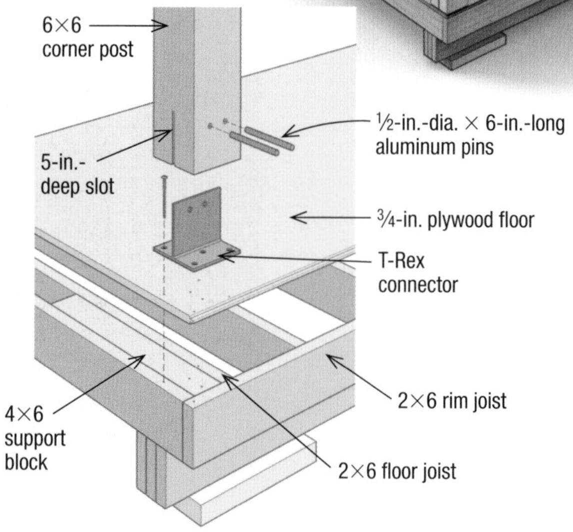
6×6 corner post
½-in.-dia. × 6-in.-long aluminum pins
5-in.-deep slot
¾-in. plywood floor
T-Rex connector
2×6 rim joist
4×6 support block
2×6 floor joist

Corner post detail

Set the 20-ft.-long skids into place atop the rows of foundation blocks. Check each skid for level, adjusting the shims, if necessary.

If your local lumberyard doesn't carry 20-ft.-long 2×6s, you'll have to order them, which may take up to a week. Another option is to splice together 8-ft.- and 12-ft.-long 2×6s to form each skid. Just be sure to stagger the splices for optimum strength.

After nailing together the skids, set them into place on top of the foundation blocks (3). Be sure to position the skids on edge, not down flat. Check that the skids rest on the center of the blocks, and that the ends of the skids extend 3 in. past the concrete blocks. Next, lay the long 2×6 and 4-ft. level perpendicularly across all three skids and check for level. Shim where necessary to ensure the three skids are level with one another.

Measure the distance between the skids, making sure they're parallel. Then, measure the two diagonal distances from opposite corners. If the two dimensions are identical, then the outer skids are square to each other. If not, adjust the first or third skid until the diagonal measurements are the same.

BUILD THE FLOOR FRAME

With the skid foundation completed, you can build the floor frame, which consists of floor joists and perimeter rim joists, all cut from pressure-treated 2×6s. Start by cutting 18 floor joists to 13 ft. 9 in. long, and two perimeter rim joists to 20 ft. long. Also cut nine 24-in.-long pieces from a pressure-treated 4×6. These blocks will provide solid fastening for the metal T-Rex connectors.

HELPFUL HINT

When adding shims to foundation blocks, never use untreated wood or plywood; the shims will eventually rot, causing the floor frame to drop. Cut shims from weather-resistant materials, such as pressure-treated wood, composite lumber, cedar or redwood, or asphalt roof shingles.

TOOL TIP

If the ground at the building site is soft, rent or buy a hand tamper and pound the area flat beneath each concrete foundation block. Compacting the soil will help keep the blocks from sinking down into the dirt. If necessary, put down a layer of gravel to support the blocks, stabilize the soil, and combat erosion.

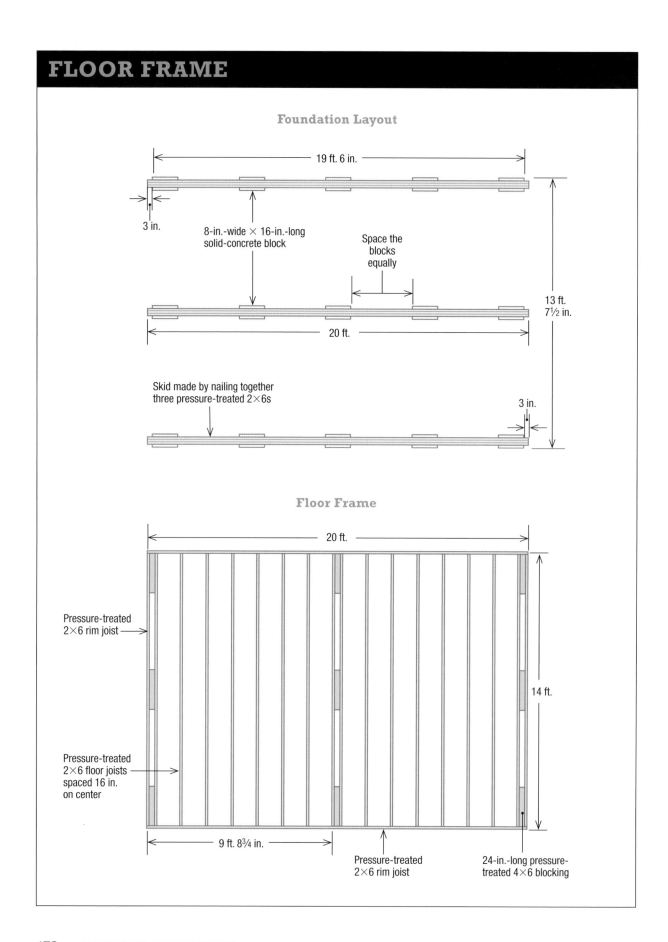

Foundation Layout

19 ft. 6 in.

3 in.

8-in.-wide × 16-in.-long solid-concrete block

Space the blocks equally

13 ft. 7½ in.

20 ft.

Skid made by nailing together three pressure-treated 2×6s

3 in.

Floor Frame

20 ft.

Pressure-treated 2×6 rim joist

Pressure-treated 2×6 floor joists spaced 16 in. on center

14 ft.

9 ft. 8¾ in.

Pressure-treated 2×6 rim joist

24-in.-long pressure-treated 4×6 blocking

Set two 2×6 joists at the far end of the skids, then place another pair in the middle. Cut two more joists for the near end of the skids.

Fasten the 2×6 joist to each of the three skids with a 3-in.-long decking screw down at an angle and into the top of the skid below.

Set two joists at each end of the skids, and two more in the middle (1). Position the three pairs of joists as shown in the Floor Frame drawing on the facing page.

Measure 9 ft. 8¾ in. from the end of each skid and draw a square line. Place the outer edge of a 2×6 joist on the line and extend its end 2½ in. past the skid. Secure the joist by driving a 3-in. decking screw down at an angle through the side of the joist and into each skid (2).

Set a 24-in.-long 4×6 block against the inside of the joist. Hold it flush with the end of the joist and drive 3-in. screws through the joist and into the 4×6 (3). Repeat to fasten two more 4×6 blocks to the same joist: Position one block flush with the opposite end of the joist, and center the other block over the middle skid.

Note that the 4×6s positioned at the ends of the joists provide solid blocking for screwing down the metal connectors. The blocks straddling the center skid are there just to prevent the joist pairs from bowing out.

Now set another joist against the opposite side of the 4×6 blocks, effectively sandwiching the 24-in.-long blocks between two joists. Fasten the joist to the blocks with 3-in. screws (4). Then screw down through the sides of the joist and into the tops of the skids. Repeat to install the remaining two pairs of joists and 4×6 blocks.

Hold a 24-in.-long support block of pressure-treated 4×6 against the inside of the joist. Drive 3-in. screws through the joist and into the block.

Place another 2×6 floor joist against the opposite side of the 4×6 block and secure it with 3-in.-long galvanized decking screws.

Slide the remaining 2×6 floor joists into place across the skids and between the double-pair joists. Space the joists 16 in. on center.

Adjust the floor joists to extend 2½ in. past the skids. Fasten each 2×6 joist by driving 3-in. screws down at an angle into the skid.

After installing the three pairs of double joists, lay out the remaining 2×6 floor joists, spacing them 16 in. on center (**5**). Before screwing each joist to the skids below, be sure its end extends 2½ in. past the skids (**6**). Now nail 2×6 rim joists to the ends of the floor joists (**7**). Drive three 3½-in. (16d) galvanized nails into the end of each joist.

Next, cover the floor frame with ¾-in. tongue-and-groove plywood. Cut the plywood so that the end of each sheet falls on the center of a joist. And stagger the seams between plywood sheets by 48 in. Fasten the plywood to the joists with 2½-in.-long decking screws spaced 10 in. to 12 in. apart (**8**).

Post-and-Beam Frame

To accurately cut the rough-sawn white-pine timbers for the post-and-beam frame you'll need to use large-capacity saws, which are available at most tool rental dealers. We used three different saws, depending on the size of the timbers. The roof rafters, ridge beam, diagonal braces, and other smaller timbers were cut with a 12-in. sliding compound-miter saw or 10-in. beam saw, which is essentially an oversized circular saw with a large-diameter blade. The larger posts and beams were cut to length with a chainsaw fixture attached to a circular saw. This fixture was also used to cut slots into the ends of the posts and beams for the metal connectors.

HELPFUL HINT

To keep a board from splitting when driving screws close to the end of a board, drill a ⁵⁄₃₂-in.- or ³⁄₁₆-in.-dia. screw-shank clearance hole through the board before driving in the screw.

TOOL TIP

A cordless drill/driver is indispensable for drilling holes and driving screws, but for pure screw-driving power use a cordless impact driver. An impact driver has four advantages over a drill/driver: It runs at faster speeds; produces much more torque or power; has less bit slippage (known as cam-out); and is shorter and more compact.

Nail the 2×6 perimeter rim joist to the ends of the floor joists. Be sure to keep the rim joist flush with the top edges of the floor joists.

Cover the floor frame with ¾-in. tongue-and-groove plywood. Fasten the plywood down to the joists with 2½-in. decking screws.

PREP THE POSTS

Six 6×6 posts support the roof—one on each corner and at the middle of the two long walls. Begin by crosscutting the six posts to 84 in. long. Then use the chainsaw fixture to cut a ¼-in.-wide by 5-in.-deep slot into both ends of each post **(1)**. To ensure the slots are cut into the exact center of the posts, we made a plywood jig, but a simple screwed-in-place 1×3 fence would work just as well. Fasten the fence to the post, so that the chainsaw aligns with the center of the post. Then guide the saw shoe along the fence to cut the slot.

Use a chainsaw attachment on a circular saw to cut 5-in.-deep slots into the 6×6 posts. The plywood jig ensures straight, accurate cuts.

SAFETY FIRST

Always wear safety goggles and a dust mask when sawing wood, especially pressure-treated wood. The chemicals used in treated wood can be particularly irritating to your eyes and lungs. And after handling treated lumber, always wash your hands before eating.

Use a router to chamfer the edges of each slot. Trimming the edges to 45° will allow the metal connectors to fit tightly into the slots.

Slide the wide flange of the T-Rex connector into the slot. Temporarily screw the metal connector to the end of the post.

After slotting both ends of all six posts, chamfer the slots with a router and 45° chamfering bit (2). The chamfer allows the metal connectors to sit flat and flush against the post ends. Insert a metal T-Rex connector into each slot (3). Temporarily fasten the connectors to the posts with two 1⁵⁄₈-in. screws. Now draw a dark pencil line across the edge of the metal connectors and onto the posts (4). Later, you'll need these "witness" marks to ensure the connectors and posts are installed back to their original positions.

To identify the precise position of the metal connector on the 6×6 post, draw a line across the edge of the connector and onto the post.

Mount a hole-drilling jig to the post with two screws and then drill two ¹⁄₂-in.-dia. holes completely through the 6×6 post and metal connector. Hold the drill as perfectly vertical as possible.

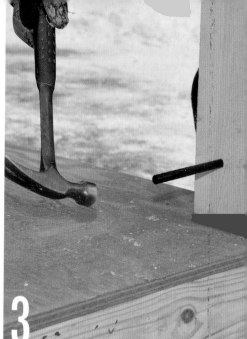

Fasten the connectors to the floor deck with four 4-in. screws. Notice that the metal flange is parallel with the long side of the barn floor.

After checking the witness marks to ensure the post is aligned properly, lift the post and set it down over the flange of the connector.

Hammer ½-in.-dia. aluminum pins through the holes to lock the 6×6 post to the connector. Tap the pins flush with the surface of the post.

Next, secure a hole-boring jig to the post with two 1⅝-in. screws. This simple jig provides a quick, accurate way to drill holes for the aluminum pins that secure the posts to the connectors. You can buy the jig or make one out of scrap wood and two ½-in.-dia. steel bushings.

With the jig in place, use an electric drill fitted with an extra-long ½-in.-dia. drill bit to bore two holes clean through both the post and connector **(5)**. Repeat to bore holes through the remaining 11 connectors.

ERECT THE POSTS

Once the metal connectors have been slotted into the posts you can install the posts, starting in one corner of the floor deck. Unscrew and remove the metal connector from the end of the post. Set the connector down onto the barn floor with its edges perfectly flush with the corner of the plywood deck. Fasten down the connector with six 4-in.-long structural screws **(1)**.

Stand the post beside the metal connector and rotate it until the witness mark on the post aligns with the mating mark on the connector. Lift the post and set it down over the protruding flange of the metal connector **(2)**. Use a hammer to drive a ½-in.-dia. by 6-in.-long aluminum pin through each hole **(3)**.

SAFETY FIRST

When drilling the ½-in.-dia. holes through the 6×6 posts and metal connectors, back out the bit several times to clear wood chips and prevent the bit from jamming and twisting your wrist. This technique also produces less strain on the drill motor.

HELPFUL HINT

Here's a quick, easy way to align the metal connectors perfectly flush with the edges of the plywood floor deck: Hold a layout square against the floor frame and allow it to protrude a couple of inches above the plywood deck. Then simply slide the connector up against the square and it'll be flush with the plywood. Don't have a layout square? Use a 10-in.- to 12-in.-long wood block instead.

Build-a-Barn Kit

All the metal connectors, building plans, and hole-boring jigs needed to erect a post-and-beam frame can be purchased through Connecticut Post and Beam (www.ctpostandbeam.com; 203-534-8771).

T-Rex connectors are available for various sizes of posts and beams, including both full-dimension rough-sawn lumber and standard nominal-dimension lumber.

Rough-sawn timbers, sometimes called "green" lumber, are cut to full dimension; a rough-sawn 6×6 post measures 6 in. by 6 in. Nominal-dimension lumber is milled and surfaced to slightly smaller sizes; a nominal 6×6 is only 5½ in. sq.; a nominal 4×4 is 3½ in. sq. Be sure to order the T-Rex connectors to match the size of your timbers.

Connecticut Post and Beam also offers more than a dozen different building plans for barns ranging in size from 12 ft. by 16 ft. to 26 ft. by 42 ft.

Temporarily hold the posts plumb with diagonal 2×4 braces. Fasten the braces to the floor frame and posts with 3-in. screws.

Install the remaining five posts in a similar manner, securing each with two pins. Then screw temporary 2×4 diagonal braces to the posts to hold them perfectly plumb (4).

INSTALL THE BEAMS

The six vertical 6×6 posts support an overhead framework of eight horizontal timbers: two 20-ft. 4-in.-long 6×10 plate beams that span the length of the barn; three 13-ft. ½-in.-long 6×10 crossbeams that run perpendicular to the plate beams; and three 6-ft. 4½-in.-long 6×6 joists that fit between two crossbeams to frame the loft.

Cut the eight beams and joists to length, then use the chainsaw attachment to cut slots into both ends of the three crossbeams and three joists; the plate beams don't require slots.

Chamfer the slots, then temporarily screw metal connectors into each slot and drill the ½-in.-dia. holes through the beams and connectors for the aluminum pins. Note that the 6×10 connectors require three holes each; the 6×6 connectors need only two holes.

Set all eight beams on edge onto the plywood floor deck. Slide one of the long plate beams up against the three posts along one side of the barn. Check that the beam extends past each corner post by 2 in. Use a framing square to mark the position of each post onto the beam.

Remove the metal T-Rex connectors from the tops of the three posts. Set the connectors onto the beam, using the post layout lines for positioning. Fasten the three connectors to the plate beam with 4-in. screws (1).

Next, raise the plate beam and set it on top of the three posts (2). You'll need at least four people to lift the heavy beam into place. Fit the

Fasten three metal connectors—one for each post—to the edge of the long 6×10 plate beam. Secure each connector with six 4-in. screws.

Lift the plate beam above the three posts. Lower the beam, guiding the metal connectors into the slots cut in the tops of the posts.

protruding flanges of the connectors into the slots in the tops of the posts. Secure the beam by driving two pins through the holes at each post (3). Repeat to install the second plate beam along the opposite side of the barn.

Prepare to install the 6×10 crossbeams by first screwing a temporary 2×6 cleat to two posts that are directly opposite one another. The cleats serve two purposes: They automatically establish the height of the crossbeam T-Rex connectors, and they hold up the crossbeams until you can drive in the aluminum pins.

Hold the 2×6 cleats level and even with the bottoms of the plate beams, not the tops of the posts. Secure each cleat with two 4-in. screws driven into the posts. Now set a T-Rex connector on top of the cleat and fasten it to the plate beam with six 4-in. screws (4). Repeat to install the connector on the opposite beam.

HELPFUL HINT

The easiest way to install the overhead beams is to first set each end of the beam on top of a stepladder. Then stand on the ladder with your shoulder pressed against the underside of the beam. Now simply take one or two steps up to set the beam into position. Depending on the size of the beam, you'll need between two and four people, each with their own ladder.

Lock the 6×10 plate beam to the 6×6 posts by driving two 1/2-in.-dia. aluminum pins through the holes at the top of each 6×6 post.

Screw a 24-in.-long temporary 2×6 cleat to the posts and then set a metal T-Rex connector on top of the cleat. Fasten the connector to the 6×10 plate beam with six 4-in.-long structural screws.

Set the 6×10 crossbeam between the plate beams. Fit the metal connectors into the slots, then rest the crossbeam on the 2×6 cleats.

Secure each end of the 6×10 crossbeam with three aluminum pins, driving the 6-in.-long pins flush with the surface of the crossbeam.

TOOL TIP

If you have trouble driving in the aluminum pins, it's because the holes in the beam and metal connector are misaligned. When that occurs, take a drill and ½-in. bit and bore into the offending holes with very light pressure to chamfer the edges of the holes in the connector. Then, when you tap in the pin, those slight chamfers will help steer the pins through the holes.

Raise the 6×10 crossbeam above the plate beams with its slots aligned vertically. Drop the beam down, fitting the metal connectors into the slots (5). Let the crossbeam rest on the 2×6 cleats. Use a hammer to drive three aluminum pins through each end of the beam, locking it between the plate beams (6). Repeat the previous four steps to install the two remaining crossbeams.

Now prepare to install three 6×6 joists between the two crossbeams at the back half of the barn. These 6×6 joists will support the floor of the overhead storage loft. Begin by screwing three metal connectors to each crossbeam. Hold the connectors flush with the tops of the crossbeams and space them 41 in. on center. Secure each connector with six 4-in. screws (7).

Screw metal connectors to the crossbeams for each of the three 6×6 joists that will support the floor of the overhead loft. Fasten a temporary 2×4 cleat to the crossbeam, directly below each metal connector.

Lower the 6×6 joist between the crossbeams. Slip the connectors into the slots in the joist ends. Rest the joist onto the cleats and then pin each end.

Install 4×6 bottom plates between the vertical 6×6 posts. Drive 8-in. screws through the plates and plywood deck and into the joists.

Fasten diagonal braces at each post to add strength and rigidity to the timber frame. Secure each 4×5 brace with four 8-in. screws.

Next, fasten a 2×4 cleat to each crossbeam, directly opposite from one another. Hold each cleat tight against the underside of the metal connector and secure it with two 3-in. screws. Lower the 6×6 joist down between the crossbeams and guide the metal flanges into the slots (8). Rest the joist on the cleats and drive two aluminum pins through each end of the joist. Repeat the previous steps to install the final two joists.

BRACE THE FRAME

Cut 4×6 bottom plates to fit snugly between the six vertical 6×6 posts (1). Secure each plate to the plywood floor deck with 8-in. structural screws. Be sure to drive the screws into a floor joist or rim joist.

Next, cut diagonal braces out of 4×5 pine timbers. Make each brace 30 in. long and miter-cut both ends to 45°. Near each end of every brace drill two ½-in.-deep by 1-in.-dia. counterbore holes. Set a brace into place between the bottom plate and post and secure with four 8-in.-long structural screws (2). After installing the lower braces, fasten the upper braces, screwing each one to the post and overhead beam (3).

Once all the braces are securely fastened, conceal the screw heads by gluing a ¼-in.-thick by 1-in.-dia. wood plug into each counterbore hole.

Add 45° diagonal braces to the upper wall sections, too. Align each brace flush with the outside edge of the post and plate beam.

Use 8-in. screws to fasten a 4×6 nailer to the underside of each 6×10 crossbeam. Hold the nailer flush with the outer surface of the beam.

Frame the sides of the window openings with 4×6 trimmer studs. Check each 4×6 for plumb with a level before screwing it in place.

FRAME THE WINDOWS

Once all the main posts and beams are erected, you can install the rough framing for the windows. There are two 24-in. by 48-in. barn-sash windows in the rear sidewall, one 24-in. by 48-in. barn-sash window in the front sidewall, and a 12-in. by 84-in. fixed (non-opening) transom window in the rear gable end wall. (The two round windows don't get framed until you frame the roof.)

However, before framing the window openings, you must install some additional wall framing. Cut two horizontal 4×6 nailers to 13 ft. long and screw one to the underside of each crossbeam at the front and rear of the barn (**1**). Position the nailers even with the 6×6 corner posts and 2 in. back from the outer face of the 6×10 crossbeam. These timbers provide nailing for the siding, and the one in the front also acts as a stop for the top of the hinged doors.

Next, create a rough opening at the front of the barn for the hinged doors: Cut two vertical 4×6 trimmer studs to span from the plywood floor to the underside of the 4×6 nailer above. Screw the trimmers in place, making sure they're plumb and spaced 84 in. apart.

Install another 4×6 trimmer stud for the sliding door on the right-hand sidewall. Cut the 4×6 to fit between the plywood floor and the underside of the 6×10 plate beam above. Screw the trimmer in place 48 in. to the right of the 6×6 center post.

HELPFUL HINT

When fastening the timbers that frame the rough openings for the doors and windows, drive in the screws from the outside of the barn. That way, the screw heads will be covered by the siding and not be visible from inside the barn.

Cut a 4×6 rough sill to fit between the vertical trimmer studs. Then push the sill down onto two 20-in.-tall spacer blocks.

Drive a 6-in. screw through the outside of the sill and into the trimmer. Move the spacer block and drive two screws up from below.

Cut 4×6 horizontal girts to span between all the posts and trimmer studs. Set the girts level with the tops of the lower diagonal braces, about 25½ in. above the floor deck. Fasten each girt with 6-in. screws. Now, frame the barn-sash windows by cutting two vertical 4×6 trimmer studs to fit between the girt and plate beam. Slip the trimmers into place, spacing them 48¼ in. apart (2). Check each trimmer for plumb, then secure them with 6-in. screws.

Saw two 4×6s to fit between the two vertical trimmer studs. One will serve as the rough sill, the other as the header. Cut two 20-in.-long spacer blocks from scrap wood. Stand one spacer on top of the horizontal girt and against the inside of each trimmer stud. Set the 4×6 sill into place between the trimmers and push it down on top of the spacer blocks (3). Screw the sill to the trimmers from the outside and from below (4).

Next, cut two 25-in.-long spacer blocks. Stand the spacers on top of the sill and against the inside of the trimmers. Set the 4×6 header between the trimmers and on top of the spacers (5). Screw the header to the trimmer studs. Remove the spacers and repeat to frame the rough openings for the remaining sidewall windows.

Saw a 4×6 header to fit between the vertical trimmer studs. Push the header down onto two 25-in.-long spacer blocks, then secure with 6-in. screws.

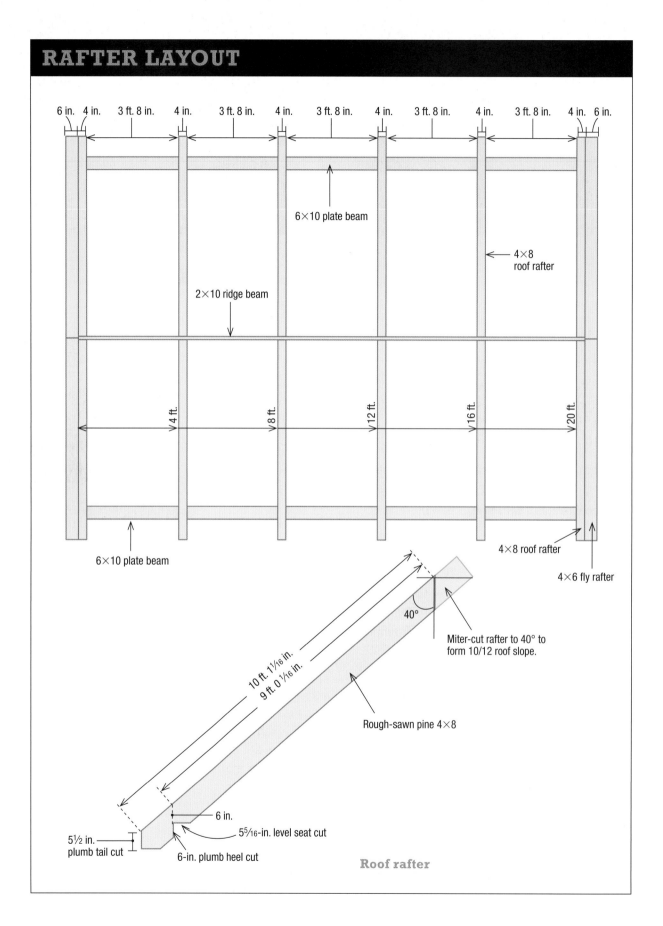

6 in. 4 in. 3 ft. 8 in. 4 in. 3 ft. 8 in. 4 in. 3 ft. 8 in. 4 in. 3 ft. 8 in. 4 in. 3 ft. 8 in. 4 in. 6 in.

6×10 plate beam

4×8 roof rafter

2×10 ridge beam

4 ft. 8 ft. 12 ft. 16 ft. 20 ft.

6×10 plate beam

4×8 roof rafter

4×6 fly rafter

40°

Miter-cut rafter to 40° to form 10/12 roof slope.

10 ft. 1¹/₁₆ in.

9 ft. 0 ¹/₁₆ in.

Rough-sawn pine 4×8

6 in.

5⁵/₁₆-in. level seat cut

5½ in. plumb tail cut

6-in. plumb heel cut

Roof rafter

Use a large portable beam saw or 12-in. sliding-compound miter saw to miter-cut the ends of the 4×8 rough-sawn roof rafters.

Hold a roof rafter in place and secure its lower end to the top of the 6×10 plate beam with two 10-in.-long structural screws.

Roof Framing

Similar to the barn frame, the roof is built with rough-sawn white-pine timbers, including 4×8 roof rafters, spaced 44 in. apart; 4×6 fly rafters, which create an overhang at each end of the gable roof; and a 2×10 ridge beam that runs between the rafters at the roof peak.

This roof is stick-built, meaning it's built one board at a time, but there are only 17 pieces to the whole roof frame—12 rafters, four 4×6 fly rafters, and one ridge beam—so it goes up pretty quickly.

SET THE RAFTERS

Cut each rafter from a 12-ft.-long 4×8. Use a 10-in. or larger beam saw or a 12-in. sliding-compound miter saw to cut the rafters. Miter the top end of each rafter to 40° to form the appropriate 10/12 roof slope (**1**). Trim the lower end of each rafter to create a 5½-in.-tall plumb tail cut, as shown in the drawing on the facing page. Then, cut the bird's-mouth notches to create a 10-in. overhang at each sidewall.

Measure 44 in. from the inside surface of the cross beams at the front and rear of the barn and mark layout lines on top of both plate beams. These lines represent the edges of the second and fifth pairs of rafters. As one person supports the upper end of the rafter, set its lower end on the line. Then, fasten the lower end of the rafter to the plate beam with two 10-in. screws (**2**). Screw on the mating rafter in a similar manner. Move down to the fifth rafter position and repeat the above steps to install the other pair of rafters.

Next, cut the 2×10 ridge beam to 20 ft. 4 in. long. Push the ridge beam up between the pairs of rafters from below (**3**). Check to be sure the ridge extends beyond each rafter pair an equal amount.

After installing two pairs of rafters, slide the 2×10 ridge beam between the rafters. Be sure the ridge is flush with the tops of the rafters and secure each rafter to the ridge with two 6-in. screws.

Create a gable overhang by attaching 4×6 fly rafters to the gable-end rafters. Fasten the fly rafters with 10-in. screws spaced 12 in. apart.

Cut 4×4s to frame the rough opening for the round windows. Check the rough sill with a 24-in. level to make sure it's perfectly level.

Screw one 4×4 brace into each corner of the rough opening. These diagonal pieces will provide support for fastening the round window.

Secure each rafter to the ridge beam with two 6-in. screws. Install the remaining four pairs of rafters in a similar manner, making sure they're spaced 44 in. apart.

Miter-cut to length four 4×6 fly rafters, two for each gable end of the roof. Hold the fly rafters against the gable-end roof rafters with their top edges flush and fasten with 10-in. screws (4).

Now frame a rough opening in each gable end for a round window. Start by installing two vertical 4×4 studs, centered beneath the ridge beam and spaced 30 in. apart. Next, cut two 4×4s to 30 in. long; make one the header, the other the rough sill. Tap the header between the vertical studs with its bottom edge 45 in. above the 6×10 cross beam. Level the header and secure it with 8-in. screws. Install the rough sill between the studs, 30 in. below the header. Check the sill for level, then screw it in place (5).

Complete the window framing by cutting four diagonal 4×4 braces to fit within the corners of the rough opening. Miter-cut both ends of each brace to 45°; make the braces 12$^{7}/_{16}$ in. long. Attach each brace with four 4-in. screws (6). Repeat the previous steps to frame a round window in the opposite gable end.

NAIL DOWN THE ROOF DECK

To prepare the roof frame for the roof shingles, you must first install a roof deck. Ordinarily, that means simply nailing down plywood sheathing and then installing the roofing. But here, we tried something a little different.

Install skipped sheathing halfway up the roof, then stop and cover the 1×8s with ½-in. plywood. Be sure to nail into the rafters.

Use roof brackets and a scaffold plank to reach the upper half of the roof. Continue to nail down 1×8s all the way to the roof peak.

To lend a more traditional look to the barn, we nailed rough-sawn 1×8 pine boards, spaced 1 in. apart, across the roof rafters. This technique, known as skipped sheathing or spaced sheathing, is typically used beneath a wood-shingle or wood-shake roof. Here, we installed the skipped sheathing and then nailed plywood on top to provide a suitable substrate for the asphalt roof shingles. Now, when the underside of the roof frame is viewed from inside the barn, it appears to be a traditional skipped-sheathed roof deck. And all of that exposed wood blends perfectly into the barn's timber frame.

The other benefit of this technique is that the roof-shingle nails won't poke through the underside of the roof deck, as they would if you installed plywood sheathing only.

Begin by fastening the 1×8s to each rafter with three 3-in. (10d) nails. Leave a 1-in. space between the rows of 1×8s and be sure every end joint falls over the center of a rafter. Install 1×8 skipped sheathing about halfway up the roof, then stop and begin installing ½-in.-thick CDX plywood roof sheathing. Fasten the plywood with 2-in. (6d) nails, making sure to drive the nails into a rafter so they don't poke through the roof deck **(1)**.

After nailing 1×8s and plywood sheathing to the lower half of the roof, install roof brackets and scaffold planks so you can easily reach the upper half of the roof. Continue to nail down 1×8s, spaced 1 in. apart, up the roof to the ridge beam **(2)**. Then cover the 1×8s with ½-in. plywood sheathing. Repeat the previous steps to install 1×8s and plywood on the opposite side of the roof.

HELPFUL HINT

When installing 1×8 skipped roof sheathing, stop occasionally and measure up to the ridge board to make sure the 1x8s are running straight and parallel, and not slanting out of level. If necessary, adjust the 1-in. space between the rows of 1×8s to get the skipped sheathing back on course.

3

Cut the 1×6 rake board longer than needed, then nail it to the fly rafter. Use a circular saw to trim the rake flush with the fascia board.

4

Nail 1×2 trim to the 1×6 rake. This narrow piece of trim isn't required, but it does create an extra shadow line along the gable ends.

Once the roof deck is completed, install the exterior roof trim, starting with the 1×6 fascia board, which runs across the rafter tails. Fasten the fascia to each rafter tail with two 3-in. (10d) galvanized nails.

Next, nail a 1×6 rake board to each fly rafter at the gable ends of the roof. Allow the rake to run long, then use a circular saw to trim it flush with the fascia (3). Now, to create a little extra shadow line along the gable ends, nail 1×2 trim to the rake boards. Hold the 1×2 flush with the upper edge of the 1×6 rake and attach it with 2½-in. (8d) galvanized nails (4).

Roofing

To complement the barn's wood timber frame and pine siding, we installed architectural-style asphalt roof shingles that are somewhat reminiscent of weathered wood. Since the building isn't heated, it wasn't necessary to first install builder's paper or any other type of underlayment. The 13¼-in.-wide by 39⅜-in.-long asphalt shingles are simply nailed down to the plywood deck with galvanized roofing nails.

INSTALL THE DRIP CAP

Before installing the roof shingles you must first protect the lower edge of the roof deck with a piece of aluminum flashing, called drip cap or drip edge. As its name implies, the flashing helps divert rainwater off the roof, but it also conceals the exposed edge of the plywood sheathing and provides rigid support for the lowest course of roof shingles.

Set the aluminum drip-cap flashing onto the roof edge, positioned so that its wide flange lays flat on the plywood deck. Secure to the roof with 1½-in.-long nails.

Aluminum drip cap is typically sold in 10-ft. lengths, and is easily cut with aviation snips. Here, we installed bright (unpainted) aluminum drip cap, but it also comes in limited paint colors, including white and brown.

Start by setting a length of drip cap onto the roof with its narrow flange hanging down over the roof edge. Align one end of the flashing flush with the plywood sheathing, then fasten it to the roof deck with 1½-in. (4d) nails. (1). Install the remaining pieces of drip cap across the edge of the roof in a similar manner, overlapping the ends by 3 in. or 4 in. When necessary, cut the drip cap to length with aviation snips.

NAIL ON THE ROOF SHINGLES

The first step to asphalt roofing is nailing down a row of starter shingles along the lower edge and ends of the roof. However, to ensure the shingles are installed perfectly straight, you must first snap two layout chalklines.

Snap the first chalkline across the plywood roof deck, 6 in. up from the lower edge of the drip cap flashing. Then, snap a line up each end of the roof—from the eave to the ridge—6 in. in from the rake boards.

Attach a row of starter shingles along the edge of the roof with 1½-in. roofing nails (2). Align the shingles flush with the chalkline and butt them end-to-end (don't overlap them). Starter shingles are 6½ in. wide, so they'll overhang the roof edge by ½ in. Note that we used a pneumatic coil roofing nailer, but you can hand-nail the shingles, as well. Just be sure to use 1½-in.-long galvanized roofing nails. Cut the last starter shingle to fit using a utility knife or manual shingle shear.

Nail starter shingles up each end of the roof (3). Again, place the shingles on the chalkline and butt them end-to-end. With starter shingles in place, you can begin nailing architectural shingles along the roof

Nail a row of starter shingles along the edge of the roof. Align the top edge of the shingles with the chalkline snapped across the roof.

Install starter shingles up each end of the roof, making sure they follow the chalkline. Secure each shingle with 1½-in. roofing nails.

Using the stair-step method to install roof shingles allows you to complete one section of roofing before moving down to the next.

To work safely on the upper half of the roof, install metal roofing brackets and a long staging plank. Nail each bracket into a rafter.

Cover the peak of the roof with overlapping ridge shingles. Be sure to place the nails where they'll be covered by the subsequent shingle.

HELPFUL HINTS

• Be sure to drive the roofing nails through the plywood and into the 1×8 skipped sheathing below. Nails driven into the 1-in. space between the 1×8s will poke through and be visible from inside the barn.

• Starter shingles might not seem all that necessary, but here are three reasons why they're important: (1) Tests have proven that starter shingles help keep roof shingles from being blown off; (2) they reduce leaks along the susceptible edges and ends of the roof; and (3) most manufacturers will increase the wind-resistance warranty coverage on the roof, if starter shingles are used.

edge. Set the first shingle into place, flush with the starter shingles at the roof's edge and end. Fasten the shingle with four nails spaced about 2 in. from each end, and 12 in. from each end. Nail down four or five shingles, butting them end-to-end, then stop and start the second course.

Cut 6 in. off the first shingle in the second course to create the proper end-joint offset from the shingles in the first course. Set the shingle into place, aligning its bottom edge with the top edge of the notched cutout in the shingle below; refer to the Roof-Shingle Detail drawing on the facing page. Nail the shingle in place, then continue the second course by nailing down three or four full-width shingles.

Cut 11 in. from the first shingle in the third course, then continue the course by nailing down a few full shingles. Trim 17 in. from the first shingle in the fourth course, nail it down, and install a couple full shingles. When you get to the fifth course, start the sequence all over again with a full-width shingle **(4)**. This roofing technique is known as the stair-step method and it allows you to work in one area for longer periods of time, rather than walking back and forth across the entire roof one course at a time.

Now return to the first course and install four or five more shingles along the roof edge. Then continue shingling the second, third, fourth, and fifth courses, stopping each one just short of the course below. Repeat this process, one section at a time, until you reach the opposite end of the roof. Cut the last shingle in each course to fit, making sure it extends ½ in. past the rake board and flush with the starter shingle below.

ROOF-SHINGLE DETAIL

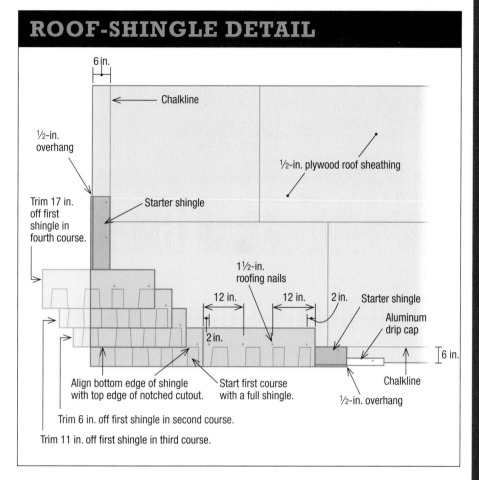

6 in.

Chalkline

½-in. overhang

½-in. plywood roof sheathing

Trim 17 in. off first shingle in fourth course.

Starter shingle

1½-in. roofing nails

12 in. 12 in. 2 in.

Starter shingle

Aluminum drip cap

2 in.

2 in.

6 in.

Align bottom edge of shingle with top edge of notched cutout.

Start first course with a full shingle.

Chalkline

½-in. overhang

Trim 6 in. off first shingle in second course.

Trim 11 in. off first shingle in third course.

Shingle Cutting Made Easy

Three-tab shingles cut easily with a sharp utility knife, but architectural shingles are a different story. Each architectural shingle is composed of two layers of thick asphalt roofing, making it very difficult to cut quickly and cleanly, especially when the shingles are cold.

On this job, we used a manual shingle shear, which operates like a giant paper cutter. Simply raise the handle, lay the shingle onto the cutting table, and bring down the handle to cut the shingle. (Be sure to keep your hand well away from the blade.) It takes quite a bit of pressure to slice through the shingle, but the shear does produce a nice, clean cut, and it'll stay sharp throughout the entire job. Shingle shears are costly to buy, but are available at most tool rental dealers.

TOOL TIP

When nailing on roof shingles, it's very important that the nail heads are driven flush. If the nails are driven too deeply or at an angle, they'll rip the shingles, increasing the chances of the shingles blowing off in a strong wind. And if the nails are left sticking up, they'll interfere with the installation of the shingles in the next course.

Once you've shingled the lower half of the roof, install roof brackets and a staging plank to shingle the upper half of the roof (5). Trim the uppermost course of shingles even with the ridge beam. Then, repeat this process to install shingles to the opposite side of the roof.

The last roofing step is to cover the peak with ridge shingles, which are simply rectangular pieces of asphalt shingle that match the architectural roof shingles. Start installing the ridge shingles at the roof end that's opposite the direction of the prevailing wind. For example, if the roof peak is aligned in an east-west direction, and the wind typically blows in from the west, then start installing the ridge shingles on the eastern end of the roof. That way, the wind will blow over the tops of the overlapping ridge shingles, forcing them down, not under them, lifting them up.

Gently bend each ridge shingle to conform to the roof peak. Be sure the shingles extend down each roof slope an equal amount, then secure each one with two nails, one per side, driven through the back half of the shingle (6). Overlap the shingles by 6 in. Secure the last ridge shingle with four nails, then seal each nail head with roofing cement.

Start installing siding in the middle of the gable end. Spline together the two center boards, so the tongue edges are facing out.

Snap a chalkline across the jagged ends of the siding. Then screw a long, straight 2×4 fence to the siding, flush on the chalkline.

Vertical-Board Siding

The barn's exterior walls are covered with rough-sawn, tongue-and-groove pine siding to match its pine timber frame. The 1x8 siding is installed vertically, in keeping with traditional barn architecture. The tongue-and-groove joints along the edges of the siding lock the boards together to seal out wind and rain.

Pine siding was also chosen because it's readily available, affordable, and attractive. However, pine isn't very weather resistant and must be protected—and then maintained—with an exterior-grade finish, such as water-repellent wood stain or paint.

Other vertical-board siding options include red cedar or redwood, two softwood species that are naturally resistant to rot and wood-boring bugs. Unfortunately, cedar and redwood sidings are typically much more expensive than pine siding.

NAIL SIDING TO THE GABLE ENDS

Install siding on the two gable ends first, then move on to the four sidewalls. Here's why: If you side the gable ends first, you won't have to worry about tools or ladders banging against and damaging the siding below.

Start siding in the very middle of a gable end, and then work your way out in both directions toward the eaves. Cut each piece of siding several inches longer than necessary, and then later, trim them all to length at the same time.

Miter-cut the upper end of a piece of siding to 40° and slide it up against the fly rafter with its tongue edge facing out toward the eave.

HELPFUL HINT

Ridge shingles get stiff and brittle when cold. (Who doesn't?) So when working in cold weather, store the ridge shingles in a heated room until they're soft and pliable. Then, when you're ready to install them, put the shingles into a bucket, bending them to fit the inside curve of the bucket. Now when you press the shingles down over the roof peak, they'll be less likely to crack or crease.

Adjust the sawblade to cut 1⅛ in. deep, then guide a circular saw along the 2×4 fence to trim the siding to length. Unscrew the 2×4.

Check it for plumb with a 4-ft. level, then face-nail the siding to the timber frame with 2½-in. spiral-shank galvanized nails. Position the nails about 1 in. in from each edge of the 1x8 siding.

Next, make a spline by ripping a ½-in.-wide strip from a sheet of ¼-in.-thick plywood. Apply waterproof glue to the spline and press it into the grooved edge on the siding piece just nailed in place. Now miter-cut a second siding piece to 40° and install it against the first siding piece, slipping its groove over the spline (1). Don't apply glue to the spline this time. Slide the second siding piece tight against the first piece, then nail it to the timber frame.

The reason for splining the first two siding pieces together groove-to-groove is that now the tongue edges are facing out in both directions, which is the preferred installation method. Continue to miter-cut and install siding to the right and then left of center, fitting the groove of one siding piece over the tongue of the previously installed piece. Again, secure the siding to the timber frame with 2½-in. galvanized nails.

Once the gable end is covered in siding, prepare to trim the siding to length. Stand at one end of the gable, near the eave, and mark a cutline onto the siding 1 in. below the bottom edge of the 6×10 crossbeam. Move to the other end of the gable and measure and mark an identical cutline. Then, take a portable circular saw and measure from the sawblade to the nearest edge of the saw shoe, a distance of about 1½ in.

Measure up from each cutline the distance from the sawblade to the shoe and draw a second set of marks, which represents the offset distance of the circular saw. Stretch a chalkline from one offset mark to the next and snap a chalkline across the siding (2).

After marking the round window opening onto the gable end, use a jigsaw to cut the 30-in.-dia. circle through the 1×8 pine siding.

Tilt the round window into the opening, then level it. Secure the window by driving four 2½-in. screws through its perimeter flange.

TOOL TIP

When installing the round windows, use a 9-in.-long torpedo level to ensure that the windows are properly positioned and not tilted out of alignment. However, since there's no place on the window on which to set the level, you must first make a setting block. Cut a ¾-in.-thick by 4-in.-long scrapwood block and set it onto one of the horizontal muntins in the center of the window grille. Now, hold the torpedo level on top of the wood block and rotate the window either clockwise or counterclockwise until the muntin is perfectly level.

Screw a long, straight 2×4 to the siding with its bottom edge flush with the chalkline. Adjust the sawblade to cut 1⅛ in. deep, then guide the saw shoe against the 2×4 fence to trim the overhanging siding pieces to length (3) (see p. 199). Unscrew the 2×4 from the wall. Repeat the previous three steps to install siding onto the gable at the opposite end of the barn.

INSTALL THE ROUND WINDOWS

To contrast the barn's straight lines and square architecture, we installed a round window into each gable-end wall. The 36-in.-dia. windows are made of cellular PVC (polyvinyl chloride), a resilient plastic that won't rot, crack, or ever need painting.

From inside the barn, locate the center of the framed rough opening for the window. Mark the center on the backside of the siding, then drive a screw through the center point to the outside. Go outside and attach a string to the tip of the protruding screw, and tie a pencil to the other end exactly 15 in. from the center point. Now swing the pencil around to strike a 30-in.-dia. circle onto the siding.

Drill a ½-in.-dia. hole through the siding, just inside the edge of the marked circle. Insert a jigsaw blade into the hole and saw along the pencil line to cut a 30-in.-dia. circle into the siding (1).

Set the round window into the opening and press it flat against the siding (2). If the window doesn't quite fit, don't force it. Use the jigsaw to remove a little more wood from the round opening, and then reinstall the window. Fasten the window by driving 2½-in.-long weather-resistant trim-head screws through the window's flange and siding and into the timber frame. Repeat this process to install the round window in the opposite gable end.

Install 1×8 pine siding across the gable-end wall. Slip each piece under the siding above and secure with 2½-in.-long galvanized nails.

If a siding piece is bowed and you can't tighten the joint, pry it closed using a wood chisel hammered into the timber frame.

NAIL SIDING TO THE WALLS

Begin siding the barn walls at the front gable end of the barn, using the same rough-sawn 1x8 pine siding installed earlier on the upper gable ends. Starting at the corner post, cut the first piece of siding to fit from the underside of the 6x10 crossbeam down 1 in. past the floor frame. This slight extension not only hides the pressure-treated rim joists but also creates a drip edge for shedding rain.

Next, use a tablesaw to rip the grooved edge off the first siding piece. Squaring up the edge is necessary because this barn doesn't have corner boards to hide the corner joints. Set the siding against the left-hand corner post with its tongue edge facing to the right and its square edge extending 1 in. past the post. This 1-in. lip will be concealed by the sidewall siding.

Use a 6-ft. level to plumb up the siding piece, then fasten it to the post with 2½-in. galvanized nails. Cut several more pieces of siding to length and nail them in place, fitting the groove of one piece over the tongue of the preceding piece **(1)**.

If a tongue-and-groove joint won't close tightly, try this trick: Start a nail in the siding piece, then hammer a wood chisel into the timber frame right beside the tongue edge of the siding. Pull on the chisel handle to pry the joint closed, then drive in the nail **(2)**.

Continue in this manner across the gable end. Cut the siding around the doorway to fit flush with the rough opening. Use the tablesaw to rip the last siding piece to width, making sure it extends 1 in. past the corner post.

Next, measure and cut the first piece of siding for the sidewall. Rip off the groove edge and set the siding piece against the corner post

3 Use a tablesaw to notch pieces of siding to fit around the rafter tails. It's often necessary to re-cut the pine siding to obtain a snug fit.

4 Stop every now and then and use a long level to check the alignment of the siding pieces above and below the window opening.

5 At the end of the window opening it's necessary to notch the siding to fit around the top, side, and bottom of the rough opening.

HELPFUL HINT

When cutting siding to fit above the window openings, cut the pieces a few inches longer than necessary. Then, notch the siding to fit around the rafter tails. Now you can re-cut the notch for a tighter fit, if necessary, without having to cut a new board. Once the notch fits tightly, mark and cut the siding piece to the finished length.

with its tongue edge facing out. Tuck the square edge behind the 1-in. siding lip protruding from the gable-end wall. Fasten the siding with 2½-in. galvanized nails. Nail up several more siding pieces across the wall. When you reach a window opening, cut the siding flush with the vertical sides of the opening and with the rough sill at the bottom of the window opening.

Cut short pieces of siding to fit above the window opening. Use a tablesaw to notch the siding, as necessary, to fit snugly around the rafter tails **(3)**. As you work your way across the window opening, stop occasionally and use a 6-ft. level to make sure the siding pieces above the window are perfectly aligned with the pieces below **(4)**.

Install a piece of siding to cover the end of the window opening. Fit it tightly against the adjacent siding pieces. Then go inside the barn and trace the window opening onto the back of the siding piece. Remove the siding, cut along the lines with a jigsaw, and set it back into place **(5)**.

Continue to install siding across the sidewall and around the corner to the rear gable end. When you get to the rough opening for the long, narrow transom window, simply nail the siding right over it **(6)**. Don't bother cutting the siding to fit around the opening. You can rout it out later when you install the transom (see p. 207). Then nail siding to the final sidewall, cutting it to fit flush around the two rough openings for the 24-in. by 48-in. windows.

Nail siding right over the transom-window opening in the rear wall. Come back later and rout out the siding from within the opening.

Doors and Windows

With the siding completed, you can install the barn doors and windows, which include two hinged doors on the front gable-end wall, a sliding door on the right-hand sidewall, and four windows: one barn sash beside the sliding door, two barn sash in the left-hand sidewall, and a long, narrow transom window in the rear gable-end wall. Start by building the doors from 1½-in.-thick pine planks.

PREP THE DOOR PLANKS

The hinged doors on the gable end and the sliding door on the sidewall are called plank doors. Each is built from 1½-in.-thick by 6¾-in.-wide tongue-and-groove pine planks. But the planks aren't simply glued together; they're reinforced with ¾-in.-dia. threaded steel rods, as shown in the drawing on p. 204. (Note that if you'd prefer to build a traditional cross-buck door, follow the instructions on p. 99 for the Board-and-Batten Shed.)

Start by cutting 12 pine planks to 84¾ in. long; that's six planks for each hinged door. Cut seven planks to 81 in. long for making the sliding barn door. Use the tablesaw to rip the grooved edge off three planks: the first plank in the left-hand hinged door, the last (sixth) plank in the right-hand hinged door, and the first plank in the sliding door.

Next, rip the tongue off the last (sixth) plank in the left-hand hinged door, the first plank in the right-hand hinged door, and the last plank in the sliding door. This ripping is necessary to create square-edged doors.

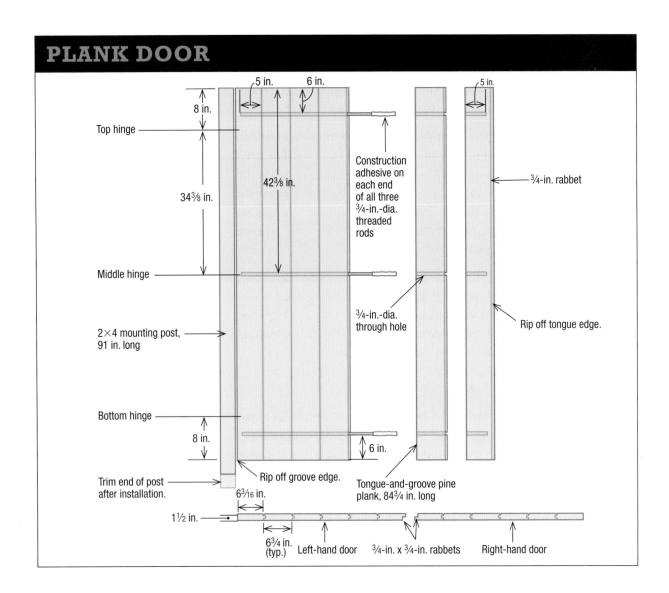

Top hinge

8 in.

5 in.   6 in.

42³⁄₈ in.

34³⁄₈ in.

Construction
adhesive on
each end
of all three
³⁄₄-in.-dia.
threaded
rods

³⁄₄-in. rabbet

Middle hinge

³⁄₄-in.-dia.
through hole

Rip off tongue edge.

2×4 mounting post,
91 in. long

Bottom hinge

8 in.

6 in.

Trim end of post
after installation.

Rip off groove edge.

6³⁄₁₆ in.

Tongue-and-groove pine
plank, 84³⁄₄ in. long

1½ in.

6³⁄₄ in.
(typ.)   Left-hand door   ³⁄₄-in. x ³⁄₄-in. rabbets   Right-hand door

TOOL TIP

If you don't own a hole-boring machine with the capacity to drill clean through the door planks, contact a professional woodworking shop or cabinet manufacturer. They'll likely have the proper equipment and may bore the holes for a nominal fee.

Now, use a tablesaw or router to cut a ³⁄₄-in.-wide by ³⁄₄-in.-deep rabbet into the face of the last plank of the left-hand hinged door, and into the rear of the first plank in the right-hand hinged door. Then, when the doors are closed, the rabbets will overlap, sealing out wind and rain.

After cutting and milling all the door planks, prepare to drill ³⁄₄-in.-dia. holes to accept the threaded rods. Each door is reinforced with three rods, so each plank receives three holes. However, the holes in the first and last planks in each door—a total of six planks—aren't bored clean through. Those six planks receive stopped holes bored to just 5 in. deep, so the threaded rods aren't visible at the door edges.

Use a drill press or horizontal boring machine to drill three ³⁄₄-in.-dia. holes into the planks. Position the holes 6 in. down from the top end of each plank, 6 in. up from the bottom end of each plank, and halfway in between at the center of the planks.

BUILD THE DOORS

Once all the holes are bored, lay out the planks for each door. Cut 6-ft.-long threaded rods to length with a hacksaw, making each about ½ in. shorter than necessary, so it doesn't bottom out in the holes.

Smear subfloor construction adhesive onto the end of a threaded rod. Spread the adhesive about 5 in. up from the rod end. Twist the glue-coated rod end into a hole in the first plank. Repeat to install the other two rods. Slide the second door plank onto the three rods—no adhesive necessary—and push its grooved edge over the tongue on the first plank. Slide on the next three planks, one at a time, tapping closed each tongue-and-groove joint.

Apply construction adhesive to the ends of the threaded rods, then press on the final plank. Use six long bar clamps or pipe clamps—three under the door, three over the door—to hold the planks together. Leave the clamps in place overnight until the adhesive dries. Repeat this procedure to assemble the two remaining doors.

HANG THE HINGED DOORS

Each hinged door swings on three 16-in.-long strap hinges. To make installation easier, attach the hinges before hanging the door. Fasten one hinge 8 in. down from the top of the door, one 8 in. up from the door bottom, and position the final hinge in the middle of the 84¾-in.-tall door.

Now, cut a 91-in.-long 2×4 mounting post and butt it against the door edge, directly underneath the hinges. Hold the upper end of the 2×4 post flush with the door top, then screw the hinges to the 2×4. Drill six ½-in.-deep by 1-in.-dia. counterbore holes into the 2×4. Position the holes a few inches above and below each hinge. Repeat these steps to mount the hinges to the remaining swinging door.

Stand the left-hand door in the doorway opening (1). Tilt in the door top until it hits the overhead 4×6 nailer. Check the door edge for plumb using a 6-ft. level. Secure the door to the barn by driving six 6-in.-long screws through the 2×4 mounting post and into the timber frame (2).

Conceal the screw heads by gluing a 1-in.-dia. wood plug into each counterbore hole (3). Use a handsaw to trim the bottom end of the 2×4 post flush with the siding. Install the right-hand door in a similar manner.

Set the door into the opening in the gable end. Note that the hinges and 2×4 mounting post are already attached to the door.

Use a cordless impact driver to drive 6-in. screws through the counterbore holes in the 2×4 mounting post and into the timber frame.

Apply waterproof glue to 1-in.-dia. hardwood plugs. Then press the plugs into the counterbore holes to conceal the screw heads.

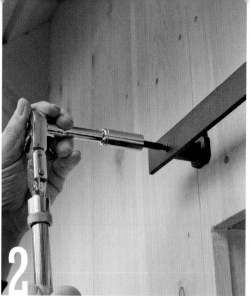

1 Fasten five iron stand-off brackets to the barn directly over the doorway opening. Mount each bracket with a single screw.

2 Attach the horizontal steel track to each stand-off bracket with a lag screw. Drive in the lags with a ratcheting socket wrench.

3 Slide the barn door onto the tracks. Each roller assembly houses a grooved wheel. Be sure the grooves fit over the top of the track.

4 To keep the sliding door from swinging away from the barn, attach a stay roller to the barn near the bottom corner of the door.

5 Cover the exposed pressure-treated rim joist in front of the sliding door with a pine board. Attach the board with 2½-in. galvanized nails.

HANG THE SLIDING DOOR

The sliding barn door hardware package consists of two U-shaped roller assemblies, a long steel track, and a door handle. Bolt the roller assemblies to the top of the door with the screws provided. Position the assemblies 6¾ in. on-center from each door edge.

Next, screw five stand-off brackets to the barn directly over the doorway opening (**1**). The brackets must be in a level, straight line. Mount the horizontal steel track to the stand-off brackets with lag screws (**2**). Be sure to drive each lag into solid framing, not just the siding.

Lean the door against the barn at the far end of the steel track. Lift the door and push it forward, aligning the grooved wheels in the roller assemblies with the track (**3**). Slide the door to the fully closed position. Install stop blocks at each end of the track to prevent the door from falling off. Slide the door back and forth a few times to ensure it rolls smoothly. If it doesn't, be sure the wheels are fully engaged with the track and that there's no debris stuck in the groove milled into the wheels.

Slide the door closed and screw a stay roller to the barn, down near the lower right-hand corner of the door (**4**). The roller will prevent the bottom of the door from swinging out away from the barn.

Now open the door and cut a rough-sawn pine board to span the width of the doorway opening. Nail the board in place to conceal the pressure-treated rim joist (**5**).

Turn on the router and then slowly move it in a clockwise direction to cut through the siding around the window's rough opening.

Install the PVC windowsill to the bottom of the opening. You could also cut the sill from cedar, redwood, or pressure-treated lumber.

INSTALL THE TRANSOM WINDOW

To cut out the opening for the fixed transom window in the rear gable-end wall, you'll need a drill with $5/8$-in.-dia. spade bit and a router fitted with a flush-trimming bit that has a ball-bearing pilot and a cutting length of $1\frac{1}{2}$ in.

Start by cutting three 1×4 pine boards to fit against the ends and top of the window's rough opening. Temporarily screw these spacer boards to the inside of the opening. The boards will allow the siding at the ends and top of the window to extend $3/4$ in. into the opening, creating stops for the window sash.

Next, from inside the barn, bore a $5/8$-in.-dia. hole through the siding near one inside corner of the rough opening. This will serve as a starter hole for the router bit. Now, go outside and adjust the router base to create a cutting depth of $1\frac{1}{8}$ in., just deep enough to cut through the 1-in.-thick siding. Place the router bit into the starter hole. Switch on the router and start routing in a clockwise direction. Guide the bit's ball-bearing pilot against the spacer boards to cut out the siding (**1**). Remove the cut siding pieces from the rough opening. Then unscrew and remove the 1×4 spacer boards.

Place the cellular PVC windowsill on top of the rough sill (**2**). Secure it with $2\frac{1}{2}$-in.-long weather-resistant trim-head screws. Slip the transom window sash into the opening and set it on top of the PVC sill (**3**). Press the sash against the siding that extends into the opening. Hold the $1/2$-in.-thick by 1-in.-wide PVC window stop tightly against the transom and then screw the stop to the header above (**4**).

Set the transom window on top of the PVC windowsill. Press the window flat against the pine siding that protrudes into the opening.

Hold the transom window in the opening by attaching a stop strip to the header above the sash. Secure the stop with trim-head screws.

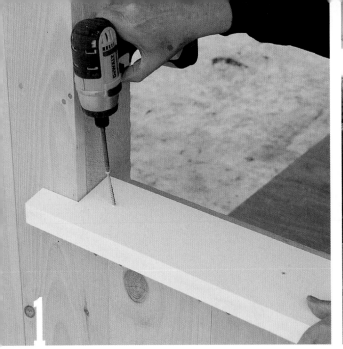

Fasten the sill into the window opening with 2½-in. trim-head screws. The horn on each end of the sill extends 3 in. past the opening.

After attaching both side casings, install the head casing across the window top where it, too, will protrude ¾ in. into the opening.

INSTALL THE BARN SASH

The barn-sash windows and trim purchased for this outbuilding are made of cellular PVC. Wood windows and trim would've cost much less, but PVC is highly weather resistant and doesn't require painting or staining. The trim package includes a sill, exterior side casings, head casing, apron, and interior stop.

Begin by setting the sill into the window opening. Fasten the PVC sill to the rough sill with 2½-in.-long trim-head screws (1). Next, screw the exterior side casings to the left and right sides of the window opening. Be sure that each casing extends ¾ in. into the opening (the window sash will close against the ¾-in. lip). Set the head casing across the top of the window, resting it on the side casings to create a ¾-in. overhang (2). Screw the head casing in place, then slip the apron under the sill and attach it with screws (3).

The barn-sash windows tilt in for ventilation. To hold them open at the proper angle, make two stops for each of the three windows. Start by cutting a 2×2 cleat to 12 in. long. Measure down 2 in. from the top end and bore a ¾-in.-dia. by ¾-in.-deep hole. Glue a ¾-in.-dia. by 1¾-in.-long wood dowel into the hole. Repeat for the remaining five sash stops.

From inside the barn, attach one stop to each side of every window opening (4). Be sure the wood dowel is facing in toward the window. Now, when you open the window, it'll come to rest against the wood dowels protruding from each cleat.

3

Install the apron under the windowsill, attaching it with 2½-in. screws. Conceal all screw heads with white caulk or plastic plugs.

4

Mount a stop to each side of the windows. When the sash is open, it'll rest against the wood dowel protruding from the 2×2 cleats.

Attach a small barrel bolt to the top interior surface of the sash. Slip the sash into place behind the stops **(5)**. Hold the sash closed and then use a hammer to lightly tap the barrel bolt to leave an impression on the header.

Remove the window and bore a 1½-in.-deep hole in the header slightly larger in diameter than the barrel bolt. Replace the sash and lock the barrel bolt. Now, set the interior stop on top of the sill and against the sash. Screw the stop to the sill to hold the bottom of the sash in place.

Slip the barn sash into place behind the wooden window stops. Install a barrel bolt at the top of the sash to hold the window closed.

5

Protection Plan

If, for some reason, you can't stain or paint the barn for several weeks, it's important to apply a coat of clear wood preservative. The preservative will help the siding repel water, dust, and dirt, and keep it from staining and rotting. Use a paint pad or paint roller to liberally apply the preservative to the siding, as shown above. Pay particular attention to the end grain of boards, which are very porous. For extra protection, apply a second coat of preservative after the first coat has dried.

Build the loft floor from 1½-in.-thick tongue-and-groove pine planks. Use a rubber mallet to pound the planks tightly together.

Create the look of a traditional rubble-rock foundation by dry-stacking thin stones around the barn. Set the stones even with the siding.

Finish Up the Barn

The last phase to completing the barn begins with installing the floor to the interior storage loft. Cut 1½-in.-thick by 6¾-in.-wide tongue-and-groove pine planks to span the 6×6 joists in the overhead loft. This is the same type of lumber used to build the plank doors. Use a rubber mallet to tap the floor planks tightly together (1). Then fasten the planks to the joists with 3½-in. nails or screws.

Next, use thin stones to build a wall around the perimeter of the barn. Tuck the stones under the barn, stacking them flush with the siding (2). The stones are dry-stacked, meaning there's no mortar holding them together. Adding stone is an optional step, but it does serve two useful purposes: It conceals the floor framing and makes the barn look as if it's sitting on a traditional stacked-stone foundation.

The final step is to finish the barn with an exterior-grade stain or paint. Here, we applied a light-gray semi-transparent stain to the siding, which protects the pine boards but still lets the wood grain show through (3). The rake boards and fascia were finished with white solid-body stain, a color chosen to complement the siding and match the windows.

Use a paintbrush to apply semi-transparent stain to the siding. Scrub the stain into the wood to ensure good adhesion and even color.

1 The rear gable-end wall has a round window and a transom window. Note the interesting shadows created by the fly rafter, rake board, and rake trim.

DESIGN DETAILS

2

3

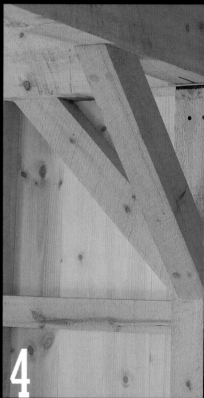

4

2. The left-hand sidewall has two large barn-sash windows that tilt in to let in fresh air. The roof is covered with architectural-style asphalt shingles.

3. The solid-pine skipped roof sheathing is visible from inside the barn. Also note how the round window fits into an octagonal rough opening.

4. The natural beauty and strength of wood is on full display in the timber frame, which consists of 6×6 posts, 6×10 beams, and 4×5 diagonal braces.

5. A wood ladder leads to a spacious overhead loft that can be used for additional storage or as a comfortable, elevated sleeping loft.

6. A sliding door permits easy access through the right-hand sidewall. The door glides on rollers, which are authentic reproductions of antique barn hardware.

5

6

RESOURCES

Building plans for sheds shown in this book

Timber-Frame Garden Shed
www.finehomebuilding.com
(800) 477-8727

Board-and-Batten Shed
Connecticut Post & Beam
www.ctpostandbeam.com
(203) 534-8771

Vinyl-Sided Storage Shed
Better Barns
www.betterbarns.com
(888) 266-1960

Cedar-Shingle Shed
Custom Plans?
Greg Butkus
gbutkus762@gmail.com
(203) 982-8530

Post-and-Beam Barn
Connecticut Post & Beam
www.ctpostandbeam.com
(203) 534-8771

DIY shed kits

Pine Harbor Wood Products
www.pineharbor.com
(800) 368-7433

Pressure-treated lumber

Arch Wood Protection
www.wolmanizedwood.com

Southern Pine Council
www.southernpine.com
(504) 443-4464

Viance Treated Wood Solutions
www.treatedwood.com

Cedar

Cedar Shake and Shingle Bureau
www.cedarbureau.org
(604) 820-7700

Western Red Cedar Lumber
Association
www.wrcla.org
(604) 684-0266

Fiber-cement siding

James Hardie Building Products
www.jameshardie.com
(888) 542-7343

Vinyl siding information

Vinyl Siding Institute
www.vinylsiding.org

PVC trim boards

Azek
www.azek.com
(877) 275-2935

CertainTeed® Corporation
www.certainteed.com

Royal Building Products
www.royalbuildingproducts.com
(800) 368-3117

Roofing and roofing products

Asphalt Roofing Manufacturers
Association
www.asphaltroofing.org

CertainTeed Corporation
www.certainteed.com

GAF® Roofing
www.gaf.com

Owens Corning®
www.owenscorning.com
(800) 438-7465

Polycarbonate roof panels

SunTuf® Panels
Palram® Americas
www.palramamericas.com
(800) 999-9459

Tuftex® Panels
Onduline North America
www.tuftexpanel.com
(800) 777-7663

Polymer slate roof tiles and cedar roof shakes

DaVinci Roofscapes
www.davinciroofscapes.com
(800) 328-4624

Gable vent

Royal Building Products
www.royalbuildingproducts.com
(800) 368-3117

Ridge and soffit vents

Cor-A-Vent
www.cor-a-vent.com
(800) 837-8368

Windows

PVC shed windows

Country Road Hardware
www.countryroadhardware.com
(203) 534-8771

Aluminum sliding windows

PGT Industries
www.pgtindustries.com
(800) 282-6019

Shed hardware

Door hinges, handles, and sliding-door hardware

Better Barns
www.betterbarns.com
(888) 266-1960

Stanley® Hardware
www.stanleyhardware.com
(800) 346-9445

Structural screws

TimberLOK® Wood Screws
www.fastenmaster.com
(800) 518-3569

T-Rex connectors

Connecticut Post & Beam
www.ctpostandbeam.com
(203) 534-8771

Miscellaneous shed products

Concrete block adhesive

Tytan® Outdoor & Landscape
High-Yield Adhesive
www.tytan.com

Self-adhesive waterproof flashing

Tite-Seal Flashing
Cofair Products
www.cofair.com
(800) 333-6700

Vinyl roll flashing

Union Corrugating Company
www.unioncorrugating.com
(888) 685-7663

INDEX

Note: Page numbers in *italics* indicate references limited to photo captions or illustrations. Page numbers in **bold** indicate resources for materials.